World Enough and Space-Time

World Enough and Space-Time

Absolute versus Relational Theories of Space and Time

John Earman

A Bradford Book
The MIT Press
Cambridge, Massachusetts
London, England

This book was set in Times New Roman by Asco Trade Typesetting Ltd., Hong Kong and printed and bound in the United States of America.

Library of Congress Cataloging-in-Publication Data

Earman, John.
World enough and space-time.
"Bradford book."
Includes index.
1. Space and time. 2. Absolute, The. 3. Relationism.
4. Science—Methodology—Philosophy. 5. Science—Philosophy. I. Title.
QC173.59.S65E17 1989 530.1′1 88-36404
ISBN 0-262-05040-4

And tear our pleasures with rough strife,
Thorough the iron grates of life.

For Frances, whose love passeth all understanding

Contents

Preface

I was pleased and flattered when Larry Sklar and Mike Friedman chose to mention me in the acknowledgments to their respective books *Space, Time, and Space-Time* and *Foundations of Space-Time Theories*. I return the compliment by stating that these are still the best available texts on a wide range of topics in the philosophy of space and time. My work may also be seen as a more substantive compliment. Sklar's and Friedman's books were especially notable for the insights and the much-needed rigor and precision they brought to the never ending struggle between absolute and relational conceptions of space and time. Now the time is ripe for achieving a fuller understanding of the dimensions and ramifications of the issues framed by Sklar and Friedman. Building on their contributions, I hope to make philosophical progress of various kinds: some of the issues can be settled; others can be sharpened; still others can be pushed aside as irrelevant; and some can be shown to break down or dissolve into the metaphysical ether. Regardless of the specifics of particular issues, the overarching goal here is to foster a better appreciation of how the absolute-relational controversy connects to problems in mathematics, physics, metaphysics, and the philosophy of scientific methodology. Foundation problems in physics, especially the general theory of relativity, are used both to advance the discussion of philosophical problems and to demonstrate that the absolute-relational controversy is not merely philosophical: it cannot be confined to the back pages of philosophy journals.

Although the treatment of some topics is necessarily technical, the organization and level of presentation of this work make it appropriate for use in an upper-level undergraduate or beginning graduate course in the philosophy of science. The bibliography, while making no pretense at completeness, is extensive enough to guide the reader into both the classic and the more up-to-date literature.

I have made no attempt to disguise my own predilections and prejudices, but at important junctions I have tried to indicate the alternative paths and the arguments pro and con for each. To some extent this may be a mistake, for philosophy might be better served if we chose simply to ignore various positions. I harbor no illusion that the considerations I marshal here achieve anything approaching closure. Indeed, I hope that this work will be judged by one of the most reliable yardsticks of fruitful philosophizing, namely, How many discussions does it engender? How many dissertation topics does it spin off?

I am grateful to the John Simon Guggenheim Memorial Foundation and to the National Science Foundation (grant no. SES-8701534) for their support of this project. Many colleagues unselfishly offered their advice on earlier drafts of this book; to one and all I offer in return my sincere appreciation, but I must especially thank Mike Friedman, Al Janis, David Malament, Tim Maudlin, John Norton, Robert Rynasiewicz, and Paul Teller. Thanks are also due to Michael Wright, whose generous support of the Σ-Club made it possible to travel to England, where I received both encouragement and helpful criticism from Harvey Brown, Jeremy Butterfield, Michael Redhead, Simon Saunders, and other members of the Oxbridge mob. Other colleagues could have, but didn't, offer help; here I would like especially to mention Larry Sklar. However, his superb selections in Chinese restaurants more than make up for this lapse.

Section 5.1 relies on "Remarks on Relational Theories of Motion," *Canadian Journal of Philosophy* 19 (1989): 83–87. Section 8.1 relies on "Locality, Non-locality, and Action at a Distance: A Skeptical Review of Some Philosophical Dogmas," in *Kelvin's Baltimore Lectures and Modern Theoretical Physics*, edited by P. Achinstein and R. Kargon (Cambridge: MIT Press, 1987). Chapter 9 relies on "What Price Space-Time Substantivalism? The Hole Story," *British Journal for the Philosophy of Science* 38 (1987): 515–525; and "Why Space Is Not a Substance (at Least Not to First Degree)," *Pacific Philosophical Quarterly* 67 (1986): 225–244. I am grateful to the editors and publishers concerned for their permissions to reuse the material here.

Notation

To make the symbolism familiar to the greatest number of readers, I have adopted the component notation for vectors and tensors instead of the newer abstract index notation. Thus, $T^{i_1 i_2 \dots i_r}_{j_1 j_2 \dots j_s}$ stands for the coordinate components of a tensor of type (r, s). The Einstein summation convention on repeated indices is in effect; e.g., $g_{ij} V^j \equiv \sum_j g_{ij} V^j$. Round and square brackets around indices are used to denote, respectively, symmetrization and antisymmetrization; e.g.,

$$T_{[ij]} \equiv \tfrac{1}{2}(T_{ij} - T_{ji}) \text{ and } T_{(ij)} \equiv \tfrac{1}{2}(T_{ij} + T_{ij}).$$

Logic

(x)	Universal quantifier
$(\exists x)$	Existential quantifier
$\rightarrow$	$P \rightarrow Q$ if P then Q
$\neg$	$\neg P$ not P

Sets

$\cup$	$A \cup B$ the union of A and B
$\cap$	$A \cap B$ the intersection of A and B
$\subset$	$A \subset B$ A is a subset of B
$\in$	$x \in A$ x is an element of A
$\varnothing$	Null set

Maps

$\phi : A \rightarrow B$	ϕ maps $x \in A$ to $\phi(x) \in B$
ϕ^{-1}	Inverse map of ϕ
$\phi \circ \psi$	Composition ψ followed by ϕ
$\phi * O$	The dragging along induced by the map ϕ of the geometric object O

Manifolds

M	An n-dimensional differentiable manifold (usually C^∞)
$\mathbb{R}$	Real line
$\mathbb{E}^n$	An n-dimensional Euclidean space
(U_α, ϕ_α)	A local coordinate chart for M determining the local coordinates x^i, i.e., $U_\alpha \subset M$ and $\phi_\alpha : M \to \mathbb{R}^n$ and $\phi_\alpha(p) = (x^1(p), x^2(p), \ldots, x^n(p))$ for $p \in M$

Derivatives and Connection

$\partial/\partial x^i$	A partial derivative with respect to the coordinate x^i
∇^2	$\partial^2/\partial x^2 + \partial^2/\partial y^2 + \partial^2/\partial z^2$
$£_V$	Lie derivative with respect to V
Γ^i_{jk}	Connection components
$\vert j$	The ordinary derivative with respect to x^j; e.g., $V^i{}_{\vert j} = \partial V^i/\partial x^j$
$\Vert j$	The covariant derivative with respect to x^j; e.g., $V^i{}_{\Vert j} = \partial V^i/\partial x^j + \Gamma^i_{jk} V^k$

Space-Time Models

$\mathfrak{M}_T$	Models of a theory T
$\langle M, O_1, O_2, \ldots \rangle$	A model consisting of a manifold M with geometric object fields O_i

World Enough and Space-Time

Introduction

Newton's *Principia*, published three centuries ago last year, launched the most successful scientific theory the world has known or is apt to know. But despite the undeniable success of the theory in explaining and unifying terrestrial and celestial motions, Newton's conceptions of space and time, conceptions that lay at the heart of his analysis of motion, were immediately and lastingly controversial. Huygens, who was arguably the greatest of Newton's Continental contemporaries, rejected Newton's absolute space and time, though he confined his misgivings to private correspondence and unpublished manuscripts. Leibniz, who was already involved in bitter priority disputes with the Newtonians, launched a public attack that culminated in the famous Leibniz–Clarke correspondence. In the eighteenth century Euler and Kant defended absolute space, though on quite different grounds, whereas Bishop Berkeley railed against it in his own inimitable fashion. In the nineteenth century Maxwell took an ambivalent stance on absolute motion, while Mach left no doubt that absolute space and time were "monstrosities" deserving no quarter. In the early twentieth century Poincaré and Einstein set their authority against the absolutists, thus helping to fashion the myth that relativity theory vindicates the relational account of space, time, and motion favored by Huygens, Leibniz, and their heirs.

Recent historical scholarship has markedly advanced our understanding of the roots of the absolute-relational controversy. Koyré and Cohen (1962) provided evidence that Newton had a hand in drafting Clarke's responses to Leibniz. (But see Hall 1980 for a contrary opinion.) The social and political context of the debate has been studied by Shapin (1981) and Stewart (1981). The origins of Newton's ideas on absolute space have been illuminated by McGuire (1978), who finds theological and kinematic motivations, as well as the better known dynamic arguments given in the *Principia.* The development of Leibniz's ideas on space and the relation of these ideas to his evolving conception of substance and his tripartite conception of reality have been explored by McGuire (1976), Winterbourne (1981), and Cover and Hartz (1986). Nevertheless, there exist a number of gaps in the literature; for instance, there is still no persuasive account of the various early reactions to Newton's "rotating bucket" experiment.

This progress in historical knowledge has, regrettably, not been matched by progress in philosophical understanding. Indeed, if progress is measured by convergence of opinion of the experts in the philosophy of space and time, then progress has been nil. In the heyday of logical positivism,

Newton's absolute space and time were seen as so much metaphysical gibberish. Reichenbach (1924), for example, cast Huygens and Leibniz in the roles of far-sighted philosophical heroes and cast Newton and Clarke in the roles of philosophical bumpkins. The postpositivist era of the 1960s and 1970s saw a major shift in attitude: absolute space was held by some to be philosophically intelligible (Lacey 1971), and in general, the issues raised by the Leibniz–Clarke debate were perceived as being more subtle than Reichenbach's caricature would suggest (Stein 1967; Earman 1970; Sklar 1976). But a deeper appreciation of the subtleties of the issues did not foster agreement. On the contrary, philosophical sympathies continued to be deeply divided, with prorelationists (van Fraassen 1970; Grünbaum 1973) matched by equally staunch proabsolutists (Nerlich 1976; Field 1980, 1985; Friedman 1983). Confident assertions that relational accounts of space and time have been or could be constructed (Suppes 1972; Bunge and Maynez 1976; Manders 1982; Mundy 1983) have been matched with skepticism about the prospects for relationism (Hooker 1971; Lacey and Anderson 1980; Butterfield 1984). Moreover, there is even disagreement over what the debate is about; some authorities see the issue of whether space or space-time is a substance as central to the debate (Sklar 1976), while others are skeptical that the issue admits a clear and interesting formulation (Malament 1976). And finally, some authors have begun to suggest that the absolute-relational dichotomy is not the best way to parse the issues and that a *tertium quid* needs to be articulated (Earman 1979; Teller 1987).

What accounts for the Tower of Babel character of the philosophical discussion? Two principal reasons suggest themselves. First, the absolute-relational controversy taps some of the most fundamental concerns in the foundations of physics, metaphysics, and scientific epistemology. This is a cause for both despair and hope: despair because if the absolute-relational controversy cannot be resolved without first settling the big questions of metaphysics and epistemology, it is not likely to be resolved, and hope because a way of making progress on the absolute-relational controversy can lead to progress on the big questions. The second and more mundane reason is that despite the fact that philosophers are professional distinction drawers, there has been a persistent failure to separate distinct though interrelated issues, issues that, contrary to the tacit assumption of many of the discussants, need not stand or fall together.

My overarching theme is that the time is ripe for progress. We are now in a position to provide a separation and classification of the issues and to see where linkages do and do not exist. In some instances we can settle an issue, at least insofar as a philosophical issue ever can be settled. And in other cases we can say with more precision what the issue is, what is at stake, and what it would take to bring a resolution. I also want to attempt to reverse a major trend in the discussion. In recent decades the absolute-relational controversy has largely become a captive of academic philosophers.[1] That the controversy is interminably debated in philosophical journals and Ph.D. dissertations is a warning sign that it has lost the relevance to contemporary science that the great natural philosophers of the seventeenth through nineteenth centuries thought that it so obviously had for the science of their day. I will attempt to correct this impression by showing, for example, how some of the very concerns raised by Leibniz in his correspondence with Clarke form the core of ongoing foundation problems in the general theory of relativity and how these problems in turn can be used to revitalize what has become an insular and bloodless philosophical discussion.

Chapter 1 provides an initial survey of the issues as they arise from Newton's *Principia* and the Leibniz–Clarke correspondence. Those already familiar with the historical source materials may be tempted to skip chapter 1, but they should be aware that the subsequent discussion relies upon the disentanglement in chapter 1 of various senses in which space and space-time can be, or fail to be, absolute. The remainder of the book divides into two roughly equal parts corresponding to the two main aspects of the absolute-relational debate. Chapters 2 to 5 deal with the twin issues of the nature of motion and the structure of space-time, while chapters 6 to 9 deal with the ontological composition of space and time (the issue of "substantivalism," to use the awkward but now well-entrenched term).

Chapter 2 describes several classical space-time structures, some of which support only relative motion, others of which ground absolute quantities of motion ranging from rotation and acceleration to velocity and spatial position. Chapter 3 surveys the epistemological, metaphysical, and empirical considerations that inform the choice of one of these structures as the structure of actual space-time. Newton thought that an adequate account of the phenomena connected with rotation could only be constructed in a space-time that provided a distinguished state of rest ("absolute space" in one of its guises). In this he was wrong, but then so were *all* of the critics, from Huygens down to Einstein, of his rotating-bucket experiment, a thesis that is defended in detail in chapter 4. Chapter 5 surveys some twentieth century developments and in particular shows why the relativistic conception of space and time is less friendly to a relational conception of motion than is the classical conception. The upshot of chapters 2 to 5 is that the relationists were wrong about the nature of motion and the structure of space and time, though they were not wrong in quite the way Newton would have it.

Chapter 6 analyzes Leibniz's famous argument against the conception of space as a substance. Although the argument undeniably has an intuitive tug, it lacks the force to dislodge the determined absolutist. Moreover, the antisubstantivalist is put on the defensive by the response that motion is absolute and that the structures needed to support absolute motion must inhere in a space or space-time distinct from bodies. Chapters 7 and 8 consider two further arguments for substantivalism, the first deriving from Kant's claim that the relationist cannot adequately account for the left-right distinction, and the second deriving from the idea that a substantival space-time is needed to support fields, which, after the special theory of relativity, are no longer regarded as states of a material medium. The trend thus seems to go resolutely against the relationist. However, the considerations of chapter 9 show that the causal version of Leibniz's argument from the principle of sufficient reason is revived by the general theory of relativity, for combining the demand for the possibility of determinism with the mutability of general-relativistic space-time structure produces a clash with the leading form of space-time substantivalism. The tentative conclusion reached is that a correct account of space and time may lie outside of the ambit of the traditional absolute-relational controversy.

In writing this book, I have attempted to compromise between making the argument continuous and cumulative and making the chapters as independent as possible. The compromise succeeds reasonably well with chapter 7 on Kant, for example. Chapters 1 to 6 help to locate Kant's precritical views on space in their historical context, but those interested in incongruent counterparts can turn directly to chapter 7 and follow the argument without absorbing the preceding chapters. The compromise is less successful in other chapters, especially the latter sections of chapter 8, which is hardly comprehensible without reference to earlier chapters, but then those sections are addressed largely to technical issues of interest only to the specialist.

I have also tried to serve both students, who need to be introduced to the issues and guided through their intricacies, and also colleagues in philosophy, who want to see some progress or at least be told something new. I am sure I will hear the charge that in trying to serve both constituencies I have served neither well. But to those students who find the going hard, I say that it is better to wrestle with difficult but exciting problems than to dine on the bland fare often served up by even the better introductory texts. Nor do I have any apologies to offer colleagues for giving an expository treatment rather than stringing together some regurgitated journal articles. They will not, I trust, find a dearth of new substantive claims in both history and philosophy of science.

1 The Origins of the Absolute-Relational Controversy

Because it conveys so vividly a sense of the once fashionable attitudes toward the absolute-relational controversy, Reichenbach's "Theory of Motion According to Newton, Leibniz, and Huygens" (1924) ought to be required reading for all students of the controversy. Here are a half dozen of Reichenbach's themes.

1. Newton was a great physicist, but both he and Samuel Clarke, his spokesman in the famous Leibniz–Clarke correspondence, were philosophical dunces.[1] ("It is ironic that Newton, who enriched science so immensely by his physical discoveries, at the same time largely hindered the development of its conceptual foundations"; Newton "turns into a mystic and a dogmatist as soon as he leaves the boundaries of his special field"; Clarke operates with "the complacency of a person not inhibited by any capacity for further enlightenment.")

2. Leibniz and Huygens were the men of real philosophical insight. (It was their "unfortunate fate to have possessed insights that were too sophisticated for the intellectual climate of their times.")

3. Newton's key mistake was to stray from his own empiricist principles. ("Newton begins with very precisely formulated empirical statements, but adds a mystical philosophical superstructure [namely, absolute space and time]."[2])

4. Newton's conception of space and time as "autonomous entities existing independently of things" shows that he was unable to emancipate himself from "the primitive notions of everyday life."

5. Leibniz's and Huygens's views on space and time are vindicated by relativity theory. ("In their opposition to Newton, physicists of our day rediscovered the answers which Newton's two contemporaries had offered in vain"; the Leibniz–Clarke correspondence "reads like a modern discussion of the theory of relativity.")

6. Newton's interpretation of the infamous rotating-bucket experiment was refuted by Mach. ("The decisive answer to Newton's argument concerning centrifugal force was given by Mach.") Mach's own interpretation of rotation is embodied in Einstein's general theory of relativity (hereafter, GTR). ("As is well known, Mach's answer is based on the fact that centrifugal force can be interpreted relativistically as a dynamic effect of gravitation produced by the rotation of the fixed stars.")

Depending upon their inclinations, students can either take heart from or stand appalled at the fact that one of the heavyweights of twentieth-century philosophy of science was so consistently wrong on so many fundamental points.[3]

1 Newton on Absolute Space and Time

Newton's Scholium on Absolute Space and Time, reproduced as an appendix to this chapter, should be read with an eye to disentangling various senses in which space, time, and space-time can be or fail to be absolute. At this stage the reader is urged to avoid making judgments about Newton's views and to concentrate instead on trying to get a firm grip on what his views are. As an initial guide, keep in mind that Newton is making at least four interrelated but distinct kinds of claims: (1) *Absolute Motion:* space and time are endowed with various structures rich enough to support an absolute, or nonrelational, conception of motion. (2) *Substantivalism:* these structures inhere in a substratum of space or space-time points. (3) *Nonconventionalism:* these structures are intrinsic to space and time. (4) *Immutability:* these structures are fixed and immutable. I pause to give the reader the opportunity to renew his acquaintance with the Scholium. (Take all the time you need; I can be patient.)

A number of comments about the opening paragraphs of the Scholium are called for here. Later chapters will examine Newton's argument from rotation, which appears in the latter part of the Scholium.

Introductory paragraph

Recall Reichenbach's claim that Leibniz had emancipated himself from the primitive notions of everyday life. Newton is here declaring his own emancipation. He is warning the reader that the terms 'time,' 'space,' 'place,' and 'motion' are not being used in their ordinary-language senses but are being given special technical meanings.

Paragraph I

Three elements of this paragraph call for emphasis and explication. First, when Newton says that absolute time "flows equably," he is not to be parsed as saying that time flows and that it flows equably. A literal notion of flow would presuppose a substratum with respect to which the flow takes place. But just as Newton rejected the idea that the points of absolute space are to be located with respect to something deeper, so he would have rejected locating the instants of time with respect to anything deeper. The phrase 'flows equably' refers not to the ontology of time but to its structure. In part, Newton is asserting that it is meaningful to ask of any two events e_1 and e_2, How much time elapses between the occurrence of e_1 and e_2? Included, of course, is the special case of simultaneous or cotemporaneous events, for which the lapse is zero. Thus, according to Newton, we have absolute simultaneity and absolute duration in that there is a unique way to partition all events into simultaneity classes, and there is an observer-independent measure of the temporal interval between nonsimultaneous events. Leibniz agreed about the absolute character of simultaneity ("*whatever exists is either simultaneous with other existences or prior or posterior*" [Loemker 1970, p. 666]), though he disagreed about the grounding of the relation of simultaneity (see section 6 below).

Second, in saying that time flows equably "without relation to anything external," Newton is asserting that the temporal interval between two events is what it is independent of what bodies are in space and how they behave.

Third, in distinguishing true mathematical time from some sensible and external measure of it, Newton is asserting that the metric of time is intrinsic to temporal intervals and that talk about the lapse of time between e_1 and e_2 is not elliptical for talk about the relation of e_1 and e_2 to the behavior of a pendulum, a quartz watch, or any other physical system. Of course, we have to use such devices in attempts to come to know what the interval is, but as Newton notes later in the Scholium, any such device may give the "wrong" answer: "It may be, that there is no such thing as equable motion, whereby time may be accurately measured. All motions may be accelerated and retarded, but the flowing of absolute time is liable to no change." Those who wish to deny Newton's intrinsicality thesis often accompany the denial with an assertion of a conventionality thesis to the effect that there is no fact of the matter about what the "correct" extrinsic metric standard is.

Paragraph II

"Absolute space, in its own nature, without relation to anything external, remains always similar and immovable." Newton is here asserting that the structure of space is absolute in that it remains the same from time to time in any physically possible world and from physically possible world to physically possible world. This immovable structure was assumed to be that of Euclidean three-space $\mathbb{E}^3$. In his early essay "De gravitatione" (c. 1668) Newton gave a theological motivation for this doctrine of the immutability of spatial structure: space is "immutable in nature, and this is because it is an emanent effect of an eternal and immutable being" (Hall and Hall 1962, p. 137). Even those of Newton's contemporaries who would have rejected the doctrine that space is an emanent effect of God would hardly have questioned the assumption of the fixity of spatial structure,[4] and over two centuries were to pass before the appearance of a successful scientific theory, Einstein's GTR, in which this assumption was dropped.

"Relative space is some movable dimension or measure of the absolute spaces." In part, this is a reiteration of Newton's intrinsicality and non-conventionalist stances: talk about the spatiotemporal separation of events e_1 and e_2 is not to be analyzed as talk about the relation between the events and extrinsic metric standards such as "rigid rods" and pendulum clocks. But the full meaning of Paragraph II cannot be appreciated without coming to grips with an ambiguity in Newton's use of 'space.' In one sense, 'space' means instantaneous space; that is, in space-time terminology, an instantaneous slice of space-time, which slice is supposed to have the character of $\mathbb{E}^3$ (see figure 1.1a). There are two other meanings of "space" that are implicit in paragraph II but emerge more explicitly in the following two paragraphs.

Paragraphs III and IV

"Place is a part of space which a body takes up, and is according to the space, either absolute or relative." "Absolute motion is the translation of a body from one absolute place into another; and relative motion, the translation from one relative place into another." A second meaning of 'space' that emerges from these passages is that of a reference frame, or a means of identifying spatial locations through time. To claim that space is absolute in this sense is to claim that there is a unique, correct way to make the identification so that for any two events e_1 and e_2, even ones lying in different instantaneous spaces (see figure 1.1b), it is meaningful to ask, Do e_1 and e_2 occur at the same spatial location? The identification procedure can be given by specifying a system of paths oblique to the planes of absolute simultaneity; with the specification indicated in figure 1.1b, the answer to the question is no.

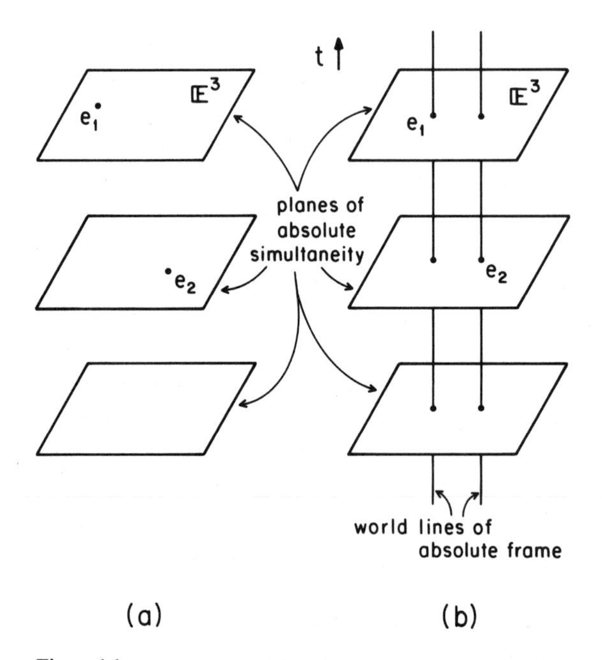

Figure 1.1
Absolute simultaneity and absolute position in Newtonian space-time

In yet a third sense, 'space' denotes a substance or substratum of points underlying physical events. The absolute frame of reference may then be thought of as being generated by the world lines of the points of absolute (or substantival) space. Spatial relations among bodies are parasitic on the spatial relations of the points of space that the bodies occupy.

Newton freely admits that "the parts of space cannot be seen, or distinguished from one another by our senses" and that as a result "in their stead we use sensible measures of them."

> And so, instead of absolute places and motions, we use relative ones; and that without any inconvenience in common affairs: but in philosophical disquisitions, we ought to abstract from our senses, and consider things themselves, distinct from what are only sensible measures of them. For it may be that there is no body really at rest, to which the places and motions of others may be referred.

A sympathetic modern gloss might run thus. Absolute space is a theoretical entity; that is, it is an entity not directly open to observation. It nevertheless makes good scientific sense to postulate this entity, because the explanation of various phenomena that are observable, particularly those involving rotation, calls for an absolute concept of motion, which in turn must be grounded on absolute space. Newton's argument here will be examined in detail in later chapters, but for now note that the argument does not involve a repudiation of the most basic form of the doctrine of the relativity of motion. Indeed, part of Newton's mistake, paradoxically, was that he joined the relational theorists in subscribing to a radical form of the relativity doctrine, according to which any assertion of the form "*x* moves" is always to be analyzed as "*x* moves relative to *y*." Where Newton parts company with the relationist is, in the first instance, in his belief that a well-founded theory of motion cannot rely only on material bodies as values for *y*.

2 Senses of Absoluteness

The following summarizes some of the leading senses of absoluteness that occur in Newton's Scholium. To mesh with the discussion in later chapters, the somewhat anachronistic but more revealing terminology of space-time is used.

1. Space-time is endowed with various structures that are intrinsic to it.

2. Among these structures are absolute simultaneity (i.e., a unique partition of events into simultaneity classes) and an absolute duration (i.e., a measure of temporal lapse that is independent of the path connecting the events).

3. There is an absolute reference frame that provides a unique way of identifying spatial locations through time. As a result, there is an absolute or well-defined measure of the velocity of individual particles and a well-defined measure of spatial separation for any pair of events.

4. The structure of space-time is immutable; i.e., it is the same from time to time in the actual world and from this world to other physically possible worlds.

5. Space-time is a substance in that it forms a substratum that underlies physical events and processes, and spatiotemporal relations among such events and processes are parasitic on the spatiotemporal relations inherent in the substratum of space-time points and regions.

Whether any or all of this is merely metaphysical gibberish, as Reichenbach would have it, remains to be seen.

3 Relationism

There are two reasons why relationism is a more elusive doctrine than absolutism. First, there is no relationist counterpart to Newton's Scholium, the *locus classicus* of absolutism. Leibniz's correspondence with Clarke is often thought to fill this role, but it falls short of articulating a coherent relational doctrine and it even fails to provide a clear account of key points in Leibniz's own version of relationism (there is, for example, no mention of Leibniz's reaction to Newton's bucket experiment). Second, there are almost as many versions of relationism as there are relationists.

At the risk of some distortion it is nevertheless useful to state three themes that form the core of classical relationism. The first theme is about both the nature of motion and the structure of space-time.

R1 All motion is the relative motion of bodies, and consequently, space-time does not have, and cannot have, structures that support absolute quantities of motion.

Huygens, as we will see in chapters 3 and 4, was a forthright exponent of this theme. Leibniz's position is more difficult to interpret, for his doctrine of "force" seems at times to threaten to undermine the relational conception of motion (see section 6.10). I shall use 'relational conception of motion' to refer to theme (R1). Please resist Reichenbach's invitation to confuse 'relational' with 'relativistic' in the sense of the special and general theories of relativity (see chapter 5).

The second theme is a denial of space-time substantivalism.

R2 Spatiotemporal relations among bodies and events are direct; that is, they are not parasitic on relations among a substratum of space points that underlie bodies or space-time points that underlie events.

This antisubstantivalist theme is sounded in Huygens's writing, especially in a number of manuscripts composed during the last years of his life and written in direct reaction to Newton's Scholium (see chapters 3 and 4). It is also sounded in Leibniz's correspondence with Clarke, as when he announces in his third letter that "As for my opinion, I have said more than once, that I hold space to be something merely relative.... For space denotes, in terms of possibility, an order of things which exist at the same time" (Alexander 1984, pp. 25–26). There follows immediately a demonstration designed to "confute the fancy of those who take space to be a substance." This famous argument will be examined in detail in chapter 6.

One of the most glaring deficiencies of the classical discussions and of the current philosophical literature is the lack of a persuasive account of the relation between (R1) and (R2). In chapter 3 I shall show that (R1) entails (R2), at least under the assumption that a minimal form of determinism is possible. Classical entries in the absolute-relational debate give the impression that both Newton and his contemporaries assumed that (R2) entails (R1), with Newton and his supporters arguing that $\neg$(R1) and, therefore by *modus tollens*, that $\neg$(R2), while relationists argued that (R2) and concluded by *modus ponens* that (R1). Unfortunately, the classical discussions are burdened by a mistake by Newton and a double countermistake by his critics. Newton seems to have thought that since "*x* moves" is short for "*x* moves relative to *y*" and since absolute motion is motion relative to absolute space, "*x* accelerates (absolutely)" means that "*x* accelerates relative to absolute space." His critics tended to swallow this move but countered correctly that Newton's sense of absolute motion entails the otiose notions of absolute velocity and absolute change of position and then concluded incorrectly that they had shown that no sense of absolute, or nonrelational, motion is required. (If the reader is confused at this juncture, I can only say, have courage and read on.) I shall argue that the failure of (R1) and other considerations do militate against (R2), although the kind of substantivalism that emerges need not be anything like that envisioned by Newton. Indeed, I shall argue that abandoning the immutability of space-time structure (as is done in GTR) while maintaining the possibility of determinism forces one to abandon a standard account of space-time substantivalism (see chapter 9). The modern upshot of the absolute-relational debate is thus a conception of space and time that is radically different from what either Newton or his critics advocated.

The third theme asserts that all spatial predication is relational in nature.

R3 No irreducible, monadic spatiotemporal properties, like 'is located at space-time point *p*,' appear in a correct analysis of the spatiotemporal idiom.[5]

While Huygens and Leibniz did not address this issue in these terms, there can be little doubt that they would have endorsed (R3), since they would have been unhappy to see smuggled in through the back door of ideology what they thought they had ruled out of the ontology with theme (R2). It is interesting to note in this regard that Leibniz's famous argument against substantivalism works equally well (or ill) against the monadic conception of spatiotemporal predication (see chapter 6).

The absolute-relational contrast is far from being a dichotomy. A possible, third alternative, which I shall call the property view of space-time, would take something from both camps: it would agree with the relationist in rejecting a substantival substratum for events while joining with the absolutist in recognizing monadic properties of spatiotemporal location. At first glance, this mongrel view does not seem to have much to recommend it, for it abandons the simplicity and parsimony that makes relationism attractive, and at the same time it gives up the ability of absolutism to explain monadic spatiotemporal properties. But like many cross breeds, this one displays a hardiness, and it will make various appearances in the chapters to come.[6]

At the risk of prejudging the outcome of future discussion, I would nevertheless like to indicate three reasons why I find Teller's (1987) version of the property view unacceptable. The first relates to his motivation. Teller claims that the substantivalist is committed to two theses: (1) that space-time points necessarily exist and (2) that necessarily, each physical event occurs at some space-time point. He concludes that "both these theses suggest ... that space-time points are abstract objects rather than concrete particulars.... So the inclination to think of space-time points as necessary suggests thinking of them as more like properties than particulars" (1987, p. 426). In response, I note that the modern substantivalist rejects thesis (1); indeed, the operation of deleting points from the space-time manifold is one of the standard devices used by general relativists in constructing cosmological models (see chapter 8). Substantivalists do accept thesis (2), but I fail to see how doing so makes space-time points analogous to quantities or determinables like mass; indeed, the substantivalist's analysis of events makes it pellucid that space-time points are being treated as substances in the sense of objects of predication. This leads me to my second reason for being unhappy with Teller's property view. I agree with Teller's sentiment that there are instances where there is no real difference between calling something a property versus calling it a concrete thing. But modern field theory does provide a powerful reason for assigning space-time points to the latter rather than the former category (see chapter 8). Finally, I will argue in chapter 9 that Einstein's "hole construction" gives a powerful reason to reject one leading form of space-time substantivalism, and exactly the same argument works against Teller's property view.

The three relational themes sounded above are largely negative: there is no absolute motion; space or space-time is not a substance; and there are no irreducible, monadic spatiotemporal properties. What, then, is the relationist's positive account of space, time, and space-time? The classical relationist can reply in two steps. First, there are at base only physical bodies, their intrinsic nonspatiotemporal properties (such as mass), and their spatiotemporal relations. Second, the relation between the absolutist and the relationist models of reality is, to use one of Leibniz's favorite concepts, one of representation, with the representation being one-many. This representational ploy will receive detailed scrutiny in the later chapters, especially chapters 6 to 9.

4 Leibniz and the Ideality of Space

In the correspondence with Clarke, Leibniz's attack on absolute space includes the charge that space and time are not fully real, that they are "ideal." In his introduction to the correspondence, H. G. Alexander states, "The ideality of space and time follows, for Leibniz, from the fact that they are neither individual substances nor aggregates of individual substances; for only these are fully real" (Alexander 1984, p. xxv). This is not a wholly satisfactory explanation. It is true that in the Leibnizian metaphysic what are ultimately real are the individual substances or monads and their nonrelational properties, and what we call the physical world is but an appearance or phenomenon.[7] But the monadology is not at issue in the polemic with Clarke, where the dispute on the nature of space and time is focused on the phenomena of physics. These phenomena are not mere appearances but are, in Leibniz's terminology, true appearances or well-founded phenomena. Indeed, there are passages from the 1680s in which Leibniz specifically refers to space and time as well-founded phenomena. ("Space, time, extension, and motion are not things but well-founded modes of our consideration." "Matter, taken for mass itself, is only a phenomenon or well founded appearance, as are space and time also."[8]) Such passages only seem to compound the puzzle of the ideality thesis.

The puzzle is resolved by noting that such passages disappear in the 1690s, when Leibniz begins to make use of a trichotomy consisting of the monads, well-founded phenomena, and a third realm consisting of entities variously labeled 'ideal', 'mental', and 'imaginary'. It is to this third category that space and time are confined in Leibniz's later writings.[9] As Cover and Hartz (1986) have emphasized, this third layer was added largely as a product of Leibniz's struggle with the labyrinth of the continuum. ("I acknowledge that time, extension, motion, and the continuum in general, as we understand them in mathematics, are only ideal things."[10]) Very roughly, Leibniz's doctrine is that in real things the part is prior to the whole, that a real thing is actually divided into definite parts; whereas in a continuum, such as space or time, the whole is prior to the parts and, indeed, there are no actual parts to a continuum but merely infinitely many potential and arbitrary divisions. ("A continuous quantity is something ideal which pertains to possibles and to actualities only in so far as they are possible. A continuum, that is, involves indeterminate parts, while on the other hand, there is nothing indefinite in actual things, in which every division is made that can be made. Actual things are compounded as is a number out of unities, ideal things as is a number out of fractions; the parts are actually in a real whole but not in the ideal whole."[11])

Leibniz forced himself to enter the labyrinth of the continuum by combining various paradoxes of infinity, learned from Galileo's writings, with a dubious reading of the axiom that the whole is greater than the part. Thus, consider the naive (and defensible) view that a continuum is actually and definitely divided, its parts being extensionless points. Leibniz rejected this conception of the composition of the continuum on the grounds that the points in, say, the interval $[0, 1]$ can be put into one-to-one correspondence with the points in a proper subinterval, say $[0, \frac{1}{2}]$, contradicting the axiom. Tracing the origins and ramifications of these quaint and unfruitful ideas is an interesting exercise in Leibniz scholarship, but it is not one that I shall attempt here.[12]

5 Other Relationisms

Modern relationists, or at least those who want to see themselves as heirs to Leibniz's philosophy of space and time, may sound one or more of the themes of section 3, but some are apt to identify relationism with the denial of Newton's claims that the metrical structures of space and time are intrinsic. Reichenbach and his followers are also intent on maintaining an ideological purity. Space and time, they hold, are constructed out of physical objects, their states, and relations between them. And further, only certain kinds of relations, taken in intension, are "objective" or "real," namely, those grounded in causal relations, such as the relation of causal connectibility. These two latter themes combine to produce various conventionality theses about space-time structure. Thus, on Reichenbach's view, the standard simultaneity relation ($\varepsilon = \frac{1}{2}$) used in the special theory of relativity (hereafter STR) is conventional unless it is definable, and perhaps uniquely definable, in terms of acceptable causal relations.[13]

The Leibniz corpus, like the Bible, can be cited in support of almost any idea, and so it is not at all surprising to find sources in Leibniz's writings for both the nonintrinsicality thesis and the causal thesis. In the "Metaphysical Foundations of Mathematics," written during the same period as the correspondence with Clarke, Leibniz clearly enunciates the nonintrinsicality thesis: "*Quantity or magnitude is that in things which can be known only through their simultaneous compresence—or by their simultaneous perception.* Thus it is impossible for us to know what a *foot* or a *yard* is unless we actually have something to serve as a measure which can be applied to successive objects after each other" (Loemker 1970, p. 667). And the same essay also contains a nascent causal theory of time: "*Time is the order of existence of things which are not simultaneous*. . . . If one of two states which are not simultaneous involves a reason for the other, the former is held to be *prior*, the latter *posterior*" (Loemker 1970, p. 666). I will have occasion to refer to those ideas at various points, but the main focus of the present study will be on the themes (R1) to (R3) of section 3.

Yet other relationist themes are to be found in the useful "Appendix on Relationism" in J. R. Lucas's *Space, Time, and Causality* (1984). However, many of the issues raised by Lucas are engaged by probing one or more of the senses of relationism already noted. Thus, for example, Lucas says that the relationist must hold as a matter of empirical fact, methodological principle, or conceptual necessity that all laws of nature are covariant (invariant?) under various sorts of transformations. In chapters 2 and 3 we shall see how invariance principles are crucial to an assessment of (R1) of section 3. As another example, Lucas's relationist holds that space and time are homogeneous and that space is isotropic. Why? One reason can be discerned by pursuing Leibniz's argument against substantivalism (theme [R2] of section 3), for the argument appears at first blush not to work if homogeneity and isotropy are abandoned (but see chapters 7 and 9). A less *ad hominem* reason is that the relationist will want to maintain that space and time are causally inefficacious. But notice that in the setting of classical space-time theories (chapter 2), space and time can be homogeneous, and space isotropic, while *space-time* is causally efficacious, because, for example, it possesses inertial or other structures that undermine the relational character of motion (contrary to theme [R1]). Another variant on the theme of the causally inefficacious status of space and time is that the very notion of causation demands it: "Causes must be repeatable: A mere difference in space and time cannot make any difference *per se*."[14] The possibility of determinism will figure in chapter 9 as part of an argument against one modern form of space-time substantivalism.

6 The Vacuum

A recurring topic in Leibniz's side of the correspondence with Clarke is criticism of the notion of a vacuum or an empty region of space.[15] The participants in this debate had the luxury of knowing what they were talking about; an empty region of space is a region unoccupied by matter. (This is the absolutist's characterization of the vacuum, but the relationist will have no trouble in providing a relational gloss, at least as long as space is not wholly empty.) We do not enjoy any such luxury for a combination of reasons: because classical particle ontology has been replaced by a dualistic particle-field ontology, because STR entails the equivalence of mass and energy, because GTR implies that the structure of space-time is not fixed and immutable, and because of the peculiarities of quantum field theory. We can maintain the spirit of the classical definition of empty space while accommodating the first two points by taking the vanishing of the relativistic energy-momentum tensor T^{ij} to be the relativistic explication of the notion of an empty space-time region. While this explication seems satisfactory in the context of STR, it has the awkward consequence of counting regions of general relativistic space-times as empty, even though these regions contain gravitational waves of sufficient strength to knock down the Rock of Gibraltar.[16] When we turn to relativistic quantum field theory, the classical notions become even more diffuse. For example, the so-called vacuum state characterizes a completely empty space (at least from the point of view of an inertial observer), but this state nonetheless contains a high degree of dynamical activity in the form of vacuum fluctuations that can have important physical consequences.[17]

But before becoming exercised at the difficulties of extending the classical notion of empty space into the relativistic and quantum realms, we should pause to consider what significance, if any, this notion has for the core of the absolute-relational controversy. A passage from Leibniz's fifth letter to Clarke seems to commit him to denying the possibility of a vacuum: "Since space in itself is an ideal thing, like time, space out of the world must be imaginary.... The case is the same with empty space within the world; which I take also to be imaginary" (Alexander 1984, p. 64). However, C. D. Broad (1946) seems to me to be on target in interpreting this passage, not as a denial of the possibility of the vacuum *per se*, but rather as a denial of the existence of substantival space, either outside of a finite material universe or inside of Guerike's vacuum pump. Earlier in the same letter Leibniz writes, "Absolutely speaking, it appears that God can make the material universe finite in extension; but the contrary appears more agreeable to his wisdom" (Alexander 1984, p. 64). I take it that Leibniz would likewise have acknowledged that God can make a universe with a vacuum inside the system of matter. The qualifier "appears" in this passage is also significant, since it is Leibniz's acknowledgment that it is not certain that the principle of sufficient reason entails that God would not actualize such a world. Indeed, all that follows from the combination of the principle of sufficient reason and the principle of plenitude is that *other things being equal*, world W_1 is better than world W_2 if W_2 contains a vacuum while W_1 does not, and therefore *other things being equal*, God would not choose to actualize W_2 over W_1. But other things might not be equal and W_2 might be preferable to W_1 because the greater simplicity and harmony of its laws outweigh its lack of plenitude. Leibniz apparently thought that such a situation is unlikely to emerge in God's preference ordering over possible worlds, but he was careful not to preclude it.

Can we then set aside the vacuum as a tangential issue, if not a complete red herring? In the most authoritative recent discussion of relationism Friedman (1983, chapter 6) thinks not, because he worries that the debate over substantivalism threatens to collapse if the world is a plenum. I will argue that his worry is misplaced, at least as regards one important form of substantivalism (see chapters 6 and 8). The real significance of the issue of the vacuum seems to me to be twofold. First, a plenum makes it easier for the relationist to maintain that absolutist models are representations of relational worlds, and second, if we look in the opposite direction, a completely empty universe is difficult for the traditional relationist to accommodate (see chapter 8).

7 Conclusion

Though cursory, our initial examination of the absolute-relational controversy affords a glimpse of how far flung, how complex, and how subtle the issues are. The glimpse also reveals what a folly it would be to wade directly into the controversy with the aim of emerging from the fray with a once-and-for-all resolution. Nevertheless, I will try to show how progress can be made by a judicious choice of lines of inquiry.[18] The most fruitful entry point, I will try to show, is theme (R1), the relational character of motion. To prepare for this entry, the next chapter is devoted to a study of various classical space-time structures.

Appendix: Newton's Scholium on Absolute Space and Time

Hitherto I have laid down the definitions of such words as are less known, and explained the sense in which I would have them to be understood in the following discourse. I do not define time, space, place, and motion, as being well known to all. Only I must observe, that the common people conceive those quantities under no other notions but from the relation they bear to sensible objects. And thence arise certain prejudices, for the removing of which it will be convenient to distinguish them into absolute and relative, true and apparent, mathematical and common.

I. Absolute, true, and mathematical time, of itself, and from its own nature, flows equably without relation to anything external, and by another name is called duration: relative, apparent, and common time, is some sensible and external (whether accurate or unequable) measure of duration by the means of motion, which is commonly used instead of true time; such as an hour, a day, a month, a year.

II. Absolute space, in its own nature, without relation to anything external, remains always similar and immovable. Relative space is some movable dimension or measure of the absolute spaces; which our senses determine by its position to bodies; and which is commonly taken for immovable space; such is the dimension of a subterraneous, an aerial, or celestial space, determined by its position in respect of the earth. Absolute and relative space are the same in figure and magnitude; but they do not remain always numerically the same. For if the earth, for instance, moves, a space of our air, which relatively and in respect of the earth remains always the same, will at one time be one part of the absolute space into which the air passes; at another time it will be another part of the same, and so, absolutely understood, it will be continually changed.

III. Place is a part of space which a body takes up, and is according to the space, either absolute or relative. I say, a part of space; not the situation, nor the external surface of the body. For the places of equal solids are always equal but their surfaces, by reason of their dissimilar figures, are often unequal. Positions properly have no quantity, nor are they so much the places themselves, as the properties of places. The motion of the whole is the same with the sum of the motions of the parts; that is, the translation of the whole, out of its place, is the same thing with the sum of the translations of the parts out of their places; and therefore the place of the whole is the same as the sum of the places as the parts, and for that reason, it is internal, and in the whole body.

IV. Absolute motion is the translation of a body from one absolute place into another; and relative motion, the translation from one relative place into another. Thus in a ship under sail, the relative place of a body is that part of the ship which the body possesses; or that part of the cavity which the body fills, and which therefore moves together with the ship: and relative rest is the continuance of the body in the same part of the ship, or of its cavity. But real, absolute rest, is the continuance of the body in the same part of that immovable space, in which the ship itself, its cavity, and all that it contains, is moved. Wherefore, if the earth is really at rest, the body, which relatively rests in the ship, will really and absolutely move with the same velocity which the ship has on the earth. But if the earth also moves, the true and absolute motion of the body will arise, partly from the true motion of the earth, in immovable space, partly from the relative motion of the ship on the earth; and if the body moves also relatively in the ship, its true motion will arise, partly from the true motion of the earth, in immovable space, and partly from the relative motions as well of the ship on the earth, as of the body in the ship; and from these relative motions will arise the relative motion of the body on the earth. As if that part of the earth, where the ship is, was truly moved towards the east, with a velocity of 10010 parts; while the ship itself, with a fresh gale, and full sails, is carried towards the west, with a velocity expressed by 10 of those parts; but a sailor walks in the ship towards the east, with 1 part of the said velocity; then the sailor will be moved truly in immovable space towards the east, with a velocity of 10001 parts, and relatively on the earth towards the west, with a velocity of 9 of those parts.

Absolute time, in astronomy, is distinguished from relative, by the equation or correction of the apparent time. For the natural days are truly unequal, though they are commonly considered as equal, and used for a measure of time; astronomers correct this inequality that they may measure the celestial motions by a more accurate time. It may be, that there is no such thing as an equable motion, whereby time may be accurately measured. All motions may be accelerated and retarded, but the flowing of absolute time is not liable to any change. The duration or perseverance of the existence of things remains the same, whether the motions are swift or slow, or none at all: and therefore this duration ought to be distinguished from what are only sensible measures thereof; and from which we deduce it, by means of the astronomical equation. The necessity of this equation, for determining the times of a phenomenon, is evinced as well from the experiments of the pendulum clock, as by eclipses of the satellites of Jupiter.

As the order of the parts of time is immutable, so also is the order of the parts of space. Suppose those parts to be moved out of their places, and they will be moved (if the expression may be allowed) out of themselves. For times and spaces are, as it were, the places as well of themselves as of all other things. All things are placed in time as to order of succession; and in space as to order of situation. It is from their essence or nature that they are places; and that the primary places of things should be movable, is absurd. These are therefore the absolute places; and translations out of those places, are the only absolute motions.

But because the parts of space cannot be seen, or distinguished from one another by our senses, therefore in their stead we use sensible measures of them. For from the positions and distances of things from any body considered as immovable, we define all places; and then with respect to such places, we estimate all motions, considering bodies as transferred from some of those places into others. And so, instead of absolute places and motions, we use relative ones; and that without any inconvenience in common affairs; but in philosophical disquisitions, we ought to abstract from our senses, and consider things themselves, distinct from what are only sensible measures of them. For it may be that there is no body really at rest, to which the places and motions of others may be referred.

But we may distinguish rest and motion, absolute and relative, one from the other by their properties, causes, and effects. It is a property of rest, that bodies really at rest do rest in respect to one another. And therefore as it is possible, that in the remote regions of the fixed stars, or perhaps far beyond them, there may be some body absolutely at rest; but impossible to know, from the position of bodies to one another in our regions, whether any of these do keep the same position to that remote body; it follows that absolute rest cannot be determined from the position of bodies in our regions.

It is a property of motion, that the parts, which retain given positions to their wholes, do partake of the motions of those wholes. For all the parts of revolving bodies endeavor to recede from the axis of motion; and the impetus of bodies moving forwards arises from the joint impetus of all the parts. Therefore, if surrounding bodies are moved, those that are relatively at rest within them will partake of their motion. Upon which account, the true and absolute motion of a body cannot be determined by the translation of it from those which only seem to rest; for the external bodies ought not only to appear at rest, but to be really at rest. For otherwise, all included bodies, besides their translation from near the surrounding ones, partake likewise of their true motions; and though that translation were not made, they would not be really at rest, but only seem to be so. For the surrounding bodies stand in the like relation to the surrounded as the exterior part of a whole does to the interior, or as the shell does to the kernel; but if the shell moves, the kernel will also move, as being part of the whole, without removal from near the shell.

A property, near akin to the preceding, is this, that if a place is moved, whatever is placed therein moves along with it; and therefore a body, which is moved from a place in motion, partakes also of the motion of its place. Upon which account, all motions, from places in motion, are no other than parts of entire and absolute motions; and every entire motion is composed of the motion of the body out of its first place, and the motion of this place out of its place; and so on, until we come to some immovable place, as in the before-mentioned example of the sailor. Wherefore, entire and absolute motions can be no otherwise determined than by immovable places; and for that reason I did before refer those absolute motions to immovable places, but relative ones to movable places. Now no other places are immovable but those that, from infinity to infinity, do all retain the same given position one to another; and upon this account must ever remain unmoved; and do thereby constitute immovable space.

The causes by which true and relative motions are distinguished, one from the other, are the forces impressed upon bodies to generate motion. True motion is neither generated nor altered, but by some force impressed upon the body moved; but relative motion may be generated or altered without any force impressed upon the body. For it is sufficient only to impress some force on other bodies with which the former is compared, that by their giving way, that relation may be changed, in which the relative rest or motion of this other body did consist. Again, true motion suffers always some change from any force impressed upon the moving body; but relative motion does not necessarily undergo any change by such forces. For if the same forces are likewise impressed on those other bodies, with which the comparison is made, that the relative position may be preserved, then that condition will be preserved in which the relative motion consists. And therefore any relative motion may be changed when the true motion remains unaltered, and the relative may be preserved when the true suffers some change. Thus, true motion by no means consists in such relations.

The effects which distinguish absolute from relative motion are, the forces of receding from the axis of circular motion. For there are no such forces in a circular motion purely relative, but in a true and absolute circular motion, they are greater or less, according to the quantity of the motion. If a vessel, hung by a long cord, is so often turned about that the cord is strongly twisted, then filled with water, and held at rest together with the water; thereupon, by the sudden action of another force, it is whirled about the contrary way, and while the cord is untwisting itself, the vessel continues for some time in this motion; the surface of the water will at first be plain, as before the vessel began to move; but after that, the vessel, by gradually communicating its motion to the water, will make it begin sensibly to revolve, and recede by little and little from the middle, and ascend to the sides of the vessel, forming itself into a concave figure (as I have experienced), and the swifter the motion becomes, the higher will the water rise, till at last, performing its revolutions in the same times with the vessel, it becomes relatively at rest in it. This ascent of the water shows its endeavor to recede from the axis of its motion; and the true and absolute circular motion of the water, which is here directly contrary to the relative, becomes known, and may be measured by this endeavor. At first, when the relative motion of the water in the vessel was greatest, it produced no endeavor to recede from the axis; the water showed no tendency to the circumference, nor any ascent towards the sides of the vessel, but remained of a plain surface, and therefore its true circular motion had not yet begun. But afterwards, when the relative motion of the water had decreased, the ascent thereof towards the sides of the vessel proved its endeavor to recede from the axis; and this endeavor showed the real circular motion of the water continually increasing, till it had acquired its greatest quantity, when the water rested relatively in the vessel. And therefore this endeavor does not depend upon any translation of the water in respect of the ambient bodies, nor can true circular motion be defined by such translation. There is only one real circular motion of any one revolving body, corresponding to only one power of endeavoring to recede from its axis of motion, as its proper and adequate effect; but relative motions, in one and the same body, are innumerable, according to the various relations it bears to external bodies, and, like other relations, are altogether destitute of any real effect, any otherwise than they may perhaps partake of that one only true motion. And therefore in their system who suppose that our heavens, revolving below the sphere of the fixed stars, carry the planets along with them; the several parts of those heavens, and the planets, which are indeed relatively at rest in their heavens, do yet really move. For they change their position one to another (which never happens to bodies truly at rest), and being carried together with their heavens, partake of their motions, and as parts of revolving wholes, endeavor to recede from the axis of their motions.

Wherefore relative quantities are not the quantities themselves, whose names they bear, but those sensible measures of them (either accurate or inaccurate), which are commonly used instead of the measured quantities themselves. And if the meaning of words is to be determined by their use, then by the names time, space, place, and motion, their [sensible] measures are properly to be understood; and the expression will be unusual, and purely mathematical, if the measured quantities themselves are meant. On this account, those violate the accuracy of language, which ought to be kept precise, who interpret these words for the measured quantities. Nor do those less defile the purity of mathematical and philosophical truths, who confound real quantities with their relations and sensible measures.

It is indeed a matter of great difficulty to discover, and effectually to distinguish, the true motions of particular bodies from the apparent; because the parts of that immovable space, in which those motions are performed, do by no means come under the observation of our senses. Yet the thing is not altogether desperate; for we have some arguments to guide us, partly from the apparent motions, which are the differences of the true motions; partly from the forces, which are the causes and effects of the true motions. For instance, if two globes, kept at a given distance one from the other by means of a cord that connects them, were revolved about their common centre of gravity, we might, from the tension of the cord, discover the endeavor of the globes to recede from the axis of their motion, and from thence we might compute the quantity of their circular motions. And then if any equal forces should be impressed at once on the alternate faces of the globes to augment or diminish their circular motions, from the increase or decrease of the tension of the cord, we might infer the increment or decrement of their motions; and thence would be found on what faces those forces ought to be impressed, that the motions of the globes might be most augmented; that is, we might discover their hindmost faces, or those which, in the circular motion, do follow. But the faces which follow being known, and consequently the opposite ones that precede, we should likewise know the determination of their motions. And thus we might find both the quantity and the determination of this circular motion, even in an immense vacuum, where there was nothing external or sensible with which the globes could be compared. But now, if in that space some remote bodies were placed that kept always a given position one to another, as the fixed stars do in our regions, we could not indeed determine from the relative translation of the globes among those bodies, whether the motion did belong to the globes or the bodies. But if we observed the cord, and found that its tension was that very tension which the motions of the globes required, we might conclude the motion to be in the globes, and the bodies to be at rest; and then, lastly, from the translation of the globes among the bodies, we should find the determination of their motions. But how we are to obtain the true motions from their causes, effects, and apparent differences, and the converse, shall be explained more at large in the following treatise. For to this end it was that I composed it. (Newton 1729, pp. 6–12)

2 Classical Space-Times

The use of modern mathematics to characterize various classical space-times is at once illuminating and dangerous. The illumination derives both from the precision afforded by the apparatus and from its power to make distinctions. But in the latter feature lies the danger. Various distinctions that we with hindsight can draw with ease were literally unavailable to the participants of the seventeenth- and eighteenth-century debates over the nature of motion and the structure of space and time. Reading these debates through the magnifying glass of this hindsight wisdom makes the participants seem at various junctures either hopelessly muddled or else lamely groping toward a seemingly obvious distinction. While such Whiggish attitudes should be avoided, no fear of being labeled Whigs should prevent us from noting that even by the standards of their own day, the positions of Newton, Leibniz, and Huygens each involved its own incoherencies. (This is in no way intended as a denigrating remark. Rather, the fact that these men and such others as Maxwell and Poincaré, a list that undeniably contains some of the most acute and penetrating intellects in the history of natural philosophy, never reached a coherent position on the absolute versus relational character of motion shows just how difficult the issues are.) Nor should the currently fashionable relativism prevent us from seeking the truth about the nature of motion and the structure of space and time. Incommensurabilities have a way of disappearing when the initially seeming incommensurable set of positions is fitted into an appropriately enlarged possibility set. The apparatus described below functions in part to provide that larger possibility set.

The reader unfamiliar with and unwilling to develop a familiarity with the mathematics used here can still profitably skim sections 1 to 7 of this chapter with an eye to answering the following: With what alternative structures can classical space-times be endowed? What absolute quantities of motion do these various structures support, and what questions about motion are meaningful in which space-times? Knowledge of the latter question can be tested by means of the quiz given in section 8. A discussion of some of the fine points of the concept of an absolute object is relegated to an appendix.

1 Machian Space-Time

This space-time and its physical applications have been studied in some detail by Barbour (1974) and Barbour and Bertotti (1977, 1982), who refer to it as Leibnizian space-time. I have chosen instead the label 'Machian space-time' both because the label 'Leibnizian space-time' has previously been applied to the space-time presented in section 2 below and because the space-time presented in this section does accurately reflect Mach's views.

The structure of Machian space-time is rather spare. It consists only of an absolute simultaneity and a Euclidean metric structure for the instantaneous spaces. To make this more precise, we start with a differentiable manifold M, assumed throughout this chapter to be the standard $\mathbb{R}^4$. M is partitioned by a family of three-dimensional hypersurfaces (planes of absolute simultaneity) that are topologically $\mathbb{R}^3$ and have the physical significance that two events are simultaneous just in case both lie on the same plane. A tangent vector at a point x, $x \in M$, is said to be timelike (respectively, spacelike) just in case it is oblique (tangent) to the plane of simultaneity passing through x. The planes of simultaneity can be identified with the level surfaces of a smooth function $t: M \to \mathbb{R}$. If time has a directionality in the minimal sense that events are ordered in time, then it is natural to choose t so as to reflect this order by requiring that for any x, $y \in M$, $t(y) > t(x)$ just in case the events at x are later than the events at y. It should be emphasized, however, that thus far t has no metrical significance, and any other time function t' such that $t' = f(t)$, $df/dt > 0$, will suffice.

Spatial distances between simultaneous events are to be well defined, and toward this end I introduce a symmetric contravariant tensor field g^{ij} of signature $(+, +, +, 0)$. Meshing g^{ij} with the simultaneity structure requires that $g^{ij}t_j = 0$, where $t_j \equiv t_{|j}$ and $|j$ denotes ordinary differentiation with respect to the jth coordinate.[1] To see that g^{ij}, which is singular and which thus does not define a space-time metric, does induce an appropriate metric on the instantaneous spaces, define a covector V_j corresponding to a spacelike vector V^i by

$$V^i = g^{ij}V_j \tag{2.1}$$

V_j is not unique, since if V_j solves (2.1), so does

$$\hat{V}_j = V_j + t_j t_k U^k. \tag{2.2}$$

for arbitrary U^k. But the spatial norms $\| V_j \|$ and $\| \hat{V}_j \|$, where $\| W_j \|^2 \equiv g^{ij} W_i W_j$, are the same. Thus, we can define the spatial norm of a spacelike vector as the norm of any of the corresponding covectors. Spatial distances are supposed to obey the laws of Euclidean geometry so that the induced space metric in the instantaneous spaces is supposed to be flat. This implies that there are coordinate systems x^i, $i = 1, 2, 3, 4$, such that $x^4 =$ constant is a plane of simultaneity and $g^{ij} = \text{diag}(1, 1, 1, 0)$. The spatial coordinates x^α, $\alpha = 1, 2, 3$, of such a coordinate system are said to be Cartesian.

The symmetries of Machian space-time are discernible by inspection. In the spatial dimension there are the usual symmetries of rigid $\mathbb{E}^3$ rotations and translations. But because we are implementing these symmetries on space-time and because Machian space-time has no preferred way to connect up the instantaneous spaces, the rotations and translations can be different on different instantaneous spaces. The space-time symmetries can be represented somewhat misleadingly in Cartesian coordinates as

$$\begin{aligned} &x^\alpha \to x'^\alpha = R^\alpha_\beta(t)x^\beta + a^\alpha(t), \qquad \alpha, \beta = 1, 2, 3 \\ &t \to t' = f(t), \qquad (df/dt) > 0 \end{aligned} \qquad \text{(Mach)}$$

where $R^\alpha_\beta(t)$ is a time dependent orthogonal matrix and the $a^\alpha(t)$ are arbitrary smooth functions of time. What is potentially misleading about this representation is that the symmetries of space-time are not coordinate transformations, in the sense of mere relabelings of space-time points by different coordinates, but rather point mappings of the space-time onto itself that preserve all of the space-time structure. Thus, as symmetry maps, the elements of (Mach) must be understood not as changes from old to new coordinates of the same points but as transformations that take one from an old point labeled (x^α, t) to a new point whose coordinates in the old coordinate system are (x'^α, t'). Despite this drawback, the coordinate representation of space-time symmetries will be used because of the convenience it affords.

The implications of symmetry transformations can be made more vivid by means of the concept of a reference frame. A global frame is specified by a congruence of smooth timelike curves.[2] Such a frame specifies a way of identifying spatial locations through time: a curve from the congruence is the world line of a point of the space of the frame, so that from the point of view of that frame two events occur at the same spatial location just in case they lie on the same curve of the congruence. Every space-time will have a preferred set of frames that reflects the structure inherent in the space-time. In Machian space-time those frames are the rigid frames, defined by the condition that the spatial distance between any two points of the frame remains constant in time.[3] The different points of a rigid frame can be labeled by Cartesian coordinates x^α, and events on the world line of a point can be labeled by t. In such an adapted coordinate system the world lines of the frame have the form $x^\alpha(t) = \text{constant}$. The first half of the transformations (Mach) can then be thought of as transformations from one rigid frame to another.

Classical world models are materialistic. Such a model is specified by giving a system of world lines of particles and their associated masses. In such a setting the absolute space-time quantities are those which are definable (at least implicitly) in terms of the space-time structure, the particle world lines, and their masses.[4] A question about the value of such a quantity is said to be meaningful. Thus, if e and f are events on particle world lines, it is always meaningful to ask in Machian space-time, Are e and f simultaneous? If the answer is no, it is not meaningful to ask, What is the spatial separation of e and f? For any two particles it is meaningful to ask, Is the spatial distance between the particles changing? If the answer is yes, it is not meaningful to ask, What is the rate of change of the distance? As we shall see in chapter 3, the meaningful/nonmeaningful distinction is one of the key elements in assessing the adequacy of a space-time as an arena for a theory of motion.

2 Leibnizian Space-Time

The label 'Leibnizian space-time' has been used by Ehlers (1973) and Stein (1977). This space-time is obtained from Machian space-time by adding a time metric. Formally, we introduce a symmetric covariant tensor field h_{ij} of signature $(0, 0, 0, +)$. A vector can be characterized as timelike (respectively, spacelike) according as $h_{ij}V^iV^j > 0$ $(=0)$. For a timelike vector, the temporal norm is defined by $\|V^i\|^2 \equiv h_{ij}V^iV^j$. To mesh with the space metric, it is required that $g^{ij}h_{ij} = 0$, which implies that there is a covector field U_i such that $h_{ij} = U_iU_j$. Some further conditions that cannot be stated at this juncture (see section 3 below) force U_i to be the gradient of a scalar field. Thus, an alternative to the above line of development would be to introduce the space and time metrics and then prove the existence of absolute time. But since we have already introduced absolute simultaneity, we will simply require at this point that U_i be the gradient of a scalar field whose level surfaces coincide with the planes of absolute simultaneity.

By choosing a t whose differences coincide with metric time differences, the symmetries of Leibnizian space-time can be represented as

$$x^\alpha \to x'^\alpha = R^\alpha_\beta(t)x^\beta + a^\alpha(t),$$
$$t \to t' = t + \text{constant}. \tag{Leib}$$

It is now meaningful to ask of two particles: How fast is the distance between them changing? How fast is the relative speed changing? etc. In general, the only questions about motion that are meaningful in this setting are questions about the relative motions of particles, which would seem to make it the ideal setting for implementing the slogan that all motion is the relative motions of bodies.

3 Maxwellian Space-Time

The label 'Maxwellian space-time' has not appeared previously in the literature. I will attempt to justify the usage in chapter 4. This space-time is obtained from Leibnizian space-time by adding a standard of rotation. Such a standard can be established by choosing a rigid frame and declaring by fiat that it is nonrotating. The family of preferred frames will then include that frame and all the other rigid frames that are nonrotating with respect to it. One would expect that the symmetries of this space-time should be like those of Leibnizian space-time except that the time dependence of the rotation matrix is killed; i.e.,

$$x^\alpha \to x'^\alpha = R^\alpha_\beta x^\beta + a^\alpha(t),$$
$$t \to t' = t + \text{constant}, \tag{Max}$$

where R^α_β is now a constant orthogonal matrix.

To describe this structure in terms of the familiar apparatus of differential geometry, we begin by looking at the class of all affine connections Γ that are compatible with the space and time metrics in that the space and time norms are preserved under parallel transport. This implies that

$$g^{ij}{}_{\|k} = h_{ij\|k} = 0, \tag{2.3}$$

where the $\|$ denotes covariant differentiation. (Condition [2.3] can be used to prove the existence of an absolute time function [see Kuchar 1981].) We now specialize to the subclass of those connections that are flat, i.e., whose

Riemann curvature tensors vanish (but see section 8 below). We then want to pick out a maximal set of such flat connections whose timelike geodesics are nonrotating with respect to one another. This can be accomplished as follows. Start with any flat $^\circ\Gamma$ satisfying (2.3). Then a congruence of timelike geodesics of $^\circ\Gamma$ defines a rigid frame. Let f^i be a spacelike vector field which is constant on the instantaneous spaces in the sense that

$$f^{i\|j} = f^i{}_{\|k} g^{kj} = 0 \tag{2.4}$$

where the $\|$ denotes covariant differentiation with respect to the chosen $^\circ\Gamma$. Then any other Γ in the set can be obtained from $^\circ\Gamma$ by

$$\Gamma^i_{jk} = {}^\circ\Gamma^i_{jk} + f^i t_j t_k \tag{2.5}$$

The reader should verify that the Γ defined in (2.4) is flat, that (2.3) holds also for covariant differentiation with respect to this Γ, that its timelike geodesics are nonrotating with respect to those of $^\circ\Gamma$, and that all rigid frames nonrotating with respect to the original are picked out in this manner. The verification can proceed by noting that a coordinate system can be chosen so that $g^{ij} = \mathrm{diag}(1,1,1,0)$; $h_{ij} = \mathrm{diag}(0,0,0,1)$; and $^\circ\Gamma^i_{jk} = 0$. The equation of geodesics of $^\circ\Gamma$ is then $d^2x^i/dt^2 = 0$. The class of all frames nonrotating with respect to these geodesics can then be generated by adding an acceleration to get $d^2x^i/dt^2 + f^i = 0$, where $f^i = (f^1(t), f^2(t), f^3(t), 0)$ and the spatial components f^α are functions of time alone. This is equivalent to shifting from $^\circ\Gamma$ to $\Gamma^i_{jk} = {}^\circ\Gamma^i_{jk} + f^i t_j t_k$, since $t_j = (0,0,0,1)$.

Now the symmetry maps must keep one within the class of preferred Γs or, equivalently, within the class of nonrotating rigid frames. It is easy to verify that these maps are represented by (Max) with the second time derivative of a^α proportional to f^α.

Although questions about the acceleration of a body are not in general meaningful in this setting, it is, of course, meaningful to ask about the state of rotation of a fluid or an extended body. Let V^i be a unit timelike vector field ($h_{ij}V^iV^j = 1$) representing the motion of a fluid or the points of a solid body. The rotation vector associated with this velocity field is

$$\Omega^i(V) \equiv \tfrac{1}{2} g^{ij} \varepsilon_{jklm} V^k g^{mn} V^l{}_{\|n}, \tag{2.6}$$

where ε_{jklm} is the natural-volume element.[5] So defined, the rotation vector is spacelike, and its spatial norm is independent of the choice of Γ from the preferred class.[6]

4 Neo-Newtonian or Galilean Space-Time

The label 'neo-Newtonian' was applied by Sklar (1976). The alternative appellation 'Galilean space-time' is justified by the form of the symmetry group, as will be seen below. This space-time is obtained from Maxwellian space-time by singling out one of the privileged affine connections, say, $*\Gamma$. Newton's Second Law for a particle of mass m is then written as

$$m\left(\frac{d^2x^i}{dt^2} + *\Gamma^i_{jk}\frac{dx^j}{dt}\frac{dx^k}{dt}\right) = F^i, \tag{2.7}$$

where F^i is the impressed force. When $F^i = 0$, (2.7) says that the world line of the particle is a geodesic of $*\Gamma$. Since $*\Gamma$ is flat, there exists a global inertial coordinate system, i.e., one in which $*\Gamma^i_{jk} = 0$, so that the inertial-force term on the left hand side of (2.7) vanishes.

The preferred frames are thus the inertial frames, frames whose world lines are straight lines in space-time. The preferred frame mappings assume the familiar Galilean form:

$$\begin{aligned} x^\alpha \to x'^\alpha &= R^\alpha_\beta x^\beta + v^\alpha t + \text{constant}, \\ t \to t' &= t + \text{constant}, \end{aligned} \tag{Gal}$$

where the v^α are constants.

It is now meaningful to ask of any given particle, regardless of whether other particles exist: Is this particle accelerating? Let V^i be the normed velocity field of the particle. Then $V^i t_i = 1$, and if we take the covariant derivative and use (2.3), we get $V^i{}_{\|j} t_i = 0$. Contracting this relation with V^j gives $a^i t_i = 0$, where $a^i \equiv V^i{}_{\|j} V^j$ is the four-vector acceleration. This shows that a^i is a spacelike vector. Thus, we can meaningfully speak of the spatial acceleration of a particle by taking the spatial norm of this vector. Similarly, contracting the equation of a geodesic, $a^i = \lambda(t)(dx^i/dt)$, with t_i shows that $\lambda(t) = 0$, i.e., t is an affine parameter, which was implictly assumed in equation (2.7).

5 Full Newtonian Space-Time

The justification for the label 'Newtonian space-time' comes from the introduction of absolute space into neo-Newtonian space-time; in effect, a particular inertial frame is singled out (see Geroch 1978; Penrose 1968; and

Sklar 1976). To do this invariantly, we introduce a unit timelike vector field A^i that is covariantly constant ($A^i{}_{\|j} = 0$). This additional structure effectively kills the velocity term in (Gal), leaving the symmetry transformations as

$$x^\alpha \to x'^\alpha = R^\alpha_\beta x^\beta + \text{constant},$$
$$t \to t' = t + \text{constant}. \qquad \text{(New)}$$

Since absolute space provides a preferred way of identifying spatial locations through time, it is now meaningful to ask of any particle: Is this particle moving? And if so, how fast is it moving? Take the tangent vector to the world line of the particle at any instant and norm it to produce the timelike unit vector W^i. The difference vector $W^i - A^i$ is spacelike, and the space norm of this vector gives the desired answer.

Notice that by using absolute space, we can define a nonsingular tensor field of signature $(+, +, +, -)$ as $\hat{g}^{ij} \equiv g^{ij} - A^i A^j$. In an inertial coordinate system adapted to absolute space, $\hat{g}^{ij} = \text{diag}(1, 1, 1, -1)$, which is reminiscent of the Minkowski metric of special relativity theory. This suggests, paradoxically, that absolute space provides a stepping stone from classical to relativistic space-times. This suggestion will be pursued in chapter 3, where $\hat{g}^{ij}$ turns out to be indispensible to formulating classical electromagnetic theory.

6 Aristotelian Space-Time

The label 'Aristotelian space-time' has been applied by Penrose (1968). This space-time is arrived at by singling out a preferred location from the absolute space of Newtonian space-time. (To be pedantic, I could introduce a velocity field C^i which is the tangent field of some particular integral curve of A^i.) To make the connection with Aristotle, we may suppose that this point corresponds to the center of the universe. Now the spatial shift term in (New) is killed and the symmetries are reduced to

$$x^\alpha \to x'^\alpha = R^\alpha_\beta x^\beta,$$
$$t \to t' = t + \text{constant}. \qquad \text{(Arist)}$$

It is now meaningful to ask of any particle, How far is this particle from the center of the universe? On any $t = \text{constant}$ plane, mark the points where the world line of the particle and the world line of the center of the universe cross the plane. The spatial distance between these two events gives the answer.

7 Other Classical Space-Times

We could extend the line of development of sections 1 to 6 and postulate a preferred direction in space and a distinguished origin of time, thus reducing the symmetry maps to the trivial identity map. Alternatively, we could expand the symmetry group of Machian space-time by dropping some element of that space-time structure, say, the Euclidean metric structure of the instantaneous spaces. The latter space-time will make a brief appearance in chapter 5, but otherwise, these alternative structures will not play any role in what follows, so there is no reason to study them further here. However, readers who are interested in learning more about the possibilities held by classical conceptions of space and time are encouraged to invent other space-times that extend the above list or fill the interstices in it.

In the usual treatment of Newtonian gravitation, gravity is an impressed force. Thus, a massive test particle moving in the gravitational field of a system of bodies feels a force F^i proportional to $g^{ij}\Phi_{|j}$, where Φ is the gravitational potential satisfying Poisson's equation $g^{ij}\Phi_{\|i\|j} = \rho$ (ρ is the mass density of the gravitational sources). An alternative approach, first suggested by Cartan, is to geometrize gravity by absorbing it into the affine structure of space-time; that is, a particle free-falling in a gravitational field no longer feels an impressed force, because it traces out a geodesic of the new Cartan connection ${}^{C}\Gamma$. Toward this end we retain condition (2.3), which expresses the compatibility of the connection with the space and time metrics, but we must drop the requirement that ${}^{C}\Gamma$ is flat and that its value is fixed *ab initio*. Poisson's equation is now replaced by $R_{ij} = \rho h_{ij}$, where R_{ij} is the Ricci tensor computed from ${}^{C}\Gamma$. The reader interested in the details of this approach may consult any number of good presentations.[7] While the details are not relevant here, it is worth emphasizing that only the unavailability of the relevant mathematical apparatus prevented the pioneers of gravitational theory from entertaining the possibility that even in the classical setting at least some elements of space-time structure do not remain "similar and immovable." It is no doubt fruitless but nonetheless tempting to speculate about whether Newton, had he possessed the relevant mathematics, would have dropped his insistence that the structure of space-time is immutable because it is an emanative effect of an immutable God in favor of the idea that the affine structure of space-time, though not immutable, is determined by laws that are immutable because they are instituted by an immutable God.

It is also natural to wonder whether the Euclidean nature of the space metric g^{ij} can be abandoned. Malament (1986) shows that the answer is no if Newtonian gravitational theory, done in the Cartan style, is to be obtained as an appropriate classical limit of general-relativistic gravitational theory.

8 Review and Quiz

Table 2.1 summarizes some of the leading features of the classical space-times studied above. As the space-time structure becomes richer, the symmetries become narrower, the list of absolute quantities increases, and more and more questions about motion become meaningful. The reader should test his knowledge of these space-times by completing the quiz, answers for which are in a note.[8]

Table 2.1
Summary of classical space-times

Name	Structures	Symmetries	Invariants
1. Machian space-time	absolute simultaneity; $\mathbb{E}^3$ structure of the instantaneous spaces	$x \to x'^{\alpha} = R^{\alpha}_{\beta}(t)x^{\beta} + a^{\alpha}(t)$ $t \to t' = f(t), \quad df/dt > 0$	relative particle distances
2. Leibnizian space-time	(1) + time metric	$x \to x'^{\alpha} = R^{\alpha}_{\beta}(t)x^{\beta} + a^{\alpha}(t)$ $t \to t' = t + \text{const}$	relative particle velocities, accelerations, etc.
3. Maxwellian space-time	(2) + standard of rotation	$x \to x' = R^{\alpha}_{\beta}x^{\beta} + a^{\alpha}(t)$ $t \to t' = t + \text{const}$	rotation for an extended body
4. neo-Newtonian space-time	(3) + inertial structure	$x \to x' = R^{\alpha}_{\beta}x^{\beta} + v^{\alpha}t + \text{const}$ $t \to t' = t + \text{const}$	acceleration of a particle
5. Full Newtonian space-time	(4) + absolute space	$x \to x' = R^{\alpha}_{\beta}x^{\beta} + \text{const}$ $t \to t' = t + \text{const}$	velocity of a particle
6. Aristotelian space-time	(5) + distinguished spatial origin	$x \to x' = R^{\alpha}_{\beta}x^{\beta}$ $t \to t' = t + \text{const}$	distance from the center of the universe

Quiz
Which questions are meaningful in which space-times?

Question	Space-time 1	2	3	4	5	6
Q1. How far is this particle from the center of the universe?	___	___	___	___	___	___
Q2. Does this particle have a speed of more than 500 m.p.h.?	___	___	___	___	___	___
Q3. Is this particle moving faster than that one?	___	___	___	___	___	___
Q4. Is this particle moving in a straight line?	___	___	___	___	___	___
Q5. Is this particle moving with constant speed?	___	___	___	___	___	___
Q6. Is the record rotating or is the turntable?	___	___	___	___	___	___
Q7. Is the rate of rotation of the record relative to the turntable changing?	___	___	___	___	___	___
Q8. Is particle #1 moving faster relative to #2 than #3 is moving relative to #4?	___	___	___	___	___	___
Q9. What is the spatial distance between event *e*, which occurs today in Pittsburgh at 3:00 P.M., and event *f*, which occurs today in N.Y.C. at 3:00 P.M.?	___	___	___	___	___	___
Q10. What is the spatial distance between event *e* (as above); and event *g*, which occurs tomorrow in N.Y.C. at 1:00 P.M.?	___	___	___	___	___	___

Appendix: Absolute Objects

Newton's dictum that absolute space remains "similar and immovable" is echoed in modern works. Thus, for example, in *A First Course in Rational Continuum Mechanics*, Truesdell speaks of a space

> the properties of which are given once and for all and are not changed by our presence or absence.... The event world is a blank canvas on which pictures of nature may be painted, the quarry for blocks from which statues of nature may be carved. This canvas, this quarry must be chosen by the artist before he sets to work. It lays the limitations upon his art, but by no means determines the pictures or the statues he will make. (1977, p. 25)

The purpose of this appendix is to expose some of the difficulties in pinpointing the notion of absolute objects, which are supposed to characterize the similar and immovable structure of the fixed space-time canvas. I shall assume that we are dealing with a space-time theory T whose intended models $\mathfrak{M}_T$ have the form $\langle M, O_1, O_2, \ldots \rangle$, where M is a differentiable manifold and the O_i are geometric object fields on M. I shall also assume that M is the same throughout $\mathfrak{M}_T$. When this assumption fails, we can either divide $\mathfrak{M}_T$ into subclasses that have the same M and apply the definition below within each subclass, or we can try to work out a localized definition of 'absolute object'. However, I suspect that in general, no very interesting definition of absolute object can be developed when M is not absolute. My strategy will be to divide the question and first to analyze the similarity part of Newton's prescription and then to turn to the immovability part.

Similarity

Let Θ be a subset of the geometric object types $\boldsymbol{O}_i$ postulated by T. We can then try to capture the similarity notion as follows.

DEFINITION Θ *remains similar* for T just in case for any $\langle M, O_1, O_2, \ldots \rangle$, $\langle M', O'_1, O'_2, \ldots \rangle \in \mathfrak{M}_T$, there is a diffeomorphism d that maps M onto M' so that $d * O_i = O'_i$ for each $\boldsymbol{O}_i \in \Theta$.[9]

For each n from 1 to 6 it is not difficult to cook up a theory T_n that incorporates the space-time from section n above and implies that the set Θ, which consists of all the object types characterizing the space-time structure, remains similar by our definition.

What does it mean for a particular object type $\boldsymbol{O}$ to remain similar relative to T? It would not be reasonable to take this to mean that the unit set $\{\boldsymbol{O}\}$ remains similar for T. For consider a theory T that demands that there is a plenum of dust particles and let $\boldsymbol{V}$ take as values smooth timelike vector fields which represent the allowed motions of the dust. Then by the proposal in question, $\boldsymbol{V}$ remains similar, whatever the details, to the laws of motion of T. But surely we will not want to count $\boldsymbol{V}$ as remaining similar if, intuitively, T allows the dust particles to perform different motions. Of course, the question facing us is precisely what standard to apply in judging similarity and difference. If the intention is to have all of the space-time structure remain similar, the obvious solution is to make the judgment relative to the fixed canvas of space-time, i.e., $\boldsymbol{V}$ is counted as remaining similar just in case $\Theta' = \{\boldsymbol{V}\} \cup \Theta$ remains similar, where Θ comprises all and only the object types characterizing the space-time structure. When Θ is sufficiently weak, similarity may emerge where we do not expect it. Thus, consider Leibnizian space-time, and suppose that the laws of motion of T require that the dust executes some form of rigid motion. Then by the proposal in question, $\boldsymbol{V}$ will be counted as remaining similar. Perhaps, however, the velocity field of the dust may not count as an absolute object, because it is not immovable, although it does remain similar.

Immovability

Consider full Newtonian space-time. The object type $\boldsymbol{A}^i$ whose values pick out the absolute frame in different models certainly remains similar if, as intended, all the other object types characterizing the space-time structure do. But now suppose that the theory T incorporating Newtonian space-time implies that the center of mass of the universe moves inertially and that the absolute frame is to be identified with the inertial frame in which the center of mass is at rest. Then we will not want to count $\boldsymbol{A}^i$ as immovable, since its values are influenced by the contingent distribution of mass ρ. More specifically, suppose that (1) $\boldsymbol{\rho}$ does not remain similar for T; i.e., there are models $\mathcal{M}$, $\mathcal{M}' \in \mathfrak{M}_T$ such that no diffeomorphism from the space-time of $\mathcal{M}$ to that of $\mathcal{M}'$ matches up all of the space-time structure (minus $\boldsymbol{A}^i$) and also matches up the values ρ and ρ' of $\boldsymbol{\rho}$ for the two models; and (2) $\boldsymbol{\rho}$ determines $\boldsymbol{A}^i$ through the laws of T; i.e., for any two models $\mathcal{M}$, $\mathcal{M}' \in \mathfrak{M}_T$, if a diffeomorphism d matches up the space-time structures (minus $\boldsymbol{A}^i$) and also the values of $\boldsymbol{\rho}$ ($d * \rho = \rho'$), then d also matches up the values of $\boldsymbol{A}^i$ ($d * A^i = A'^i$). And more generally, we would not want to count

O as immovable for *T* if there is some other object type (or set of object types) that determines ***O*** and does not remain similar for *T*. But the negation of this circumstance does not provide a sufficient condition for immovability, since presumably there are many ways ***O*** can be influenced by other objects without being determined by them.

Despite being fuzzy at the boundaries, the concept of an absolute object remains a useful one in analyzing space-time theories, and it will be employed in the chapters below.

3 Choosing a Classical Space-Time

What considerations would serve to establish that in fact space-time has a structure like one of those studied in chapter 2 or to the contrary that the structure is different from anything envisioned there? That is not a question that the participants in the seventeenth- and eighteenth-century version of the absolute-relational controversy asked or were even in a position to ask. Nevertheless, I will pursue this question, because despite its anachronistic character, it can, if asked judiciously, enhance our understanding of the historical and conceptual issues.

1 Arguments from the Meaning of 'Motion'

The general form of the argument from the meaning of 'motion' runs thus: From the very meaning of 'motion' it follows that various questions about motion are meaningful (alternatively, are not meaningful); therefore, space-time should have at least as much structure as is needed to make these questions meaningful (alternatively, space-time should not have so much structure as to make these questions meaningful).

Newton's "De gravitatione" can be taken to contain an argument to the effect that space-time has at least as much structure as full Newtonian space-time (section 2.5), because the very meaning of motion requires a determinate velocity. The first part of the argument emerges from Newton's critique of Descartes's analysis of motion. "That the absurdity of this position [Descartes's] may be disclosed in full measure, I say that thence it follows that a moving body has no determinate velocity and no definite line in which is moves.... On the contrary, there cannot be motion without a certain velocity and determination" (Hall and Hall 1962, p. 129). The argument is completed by noting that absolute space is needed to ground such a determination.

Descartes's and Huygens's conceptions lead in the opposite direction: space-time has no more structure than Machian space-time (section 2.1) or perhaps Leibnizian space-time (section 2.2), because the only talk about motion that is meaningful is talk about the relative motion of bodies. In the *Principles*, Descartes defines the motion peculiar to each body as the "transference of one part of matter or of one body, from the vicinity of those bodies immediately contiguous to it and considered at rest, into the vicinity of [some] others" (Descartes 1644, II.25).[1] There are other motions through "participation" (II.31), but these are not motions in the philosophical sense (III.29). And in any case they are all relative bodily motions.

In a manuscript that was probably composed in the early 1690s, Huygens, who had broken with Descartes over the correct form of the laws of impact, continued faithful to the Cartesian analysis of the meaning of 'motion'. "To those who ask what motion is, only this answer suggests itself: that bodies can be said to move when their place ['situs'] and their distances change, either with respect to each other or with respect to another body.... We don't understand anything more than this for motion" (Huygens 1888–1950, 16 : 227–228). A similar sentiment is found in another manuscript fragment from the same period. "Place ['locus'] indeed can only be defined or designated through other bodies. Thus, there can be no motion or rest in bodies except with respect to each other" (21 : 507).[2]

There are two ways to interpret such meaning claims. First, they can be taken to be claims about the meanings of words. Read in this way, they are wholly unpersuasive as answers to our question, for no amount of linguistic analysis can settle questions about the spatiotemporal structure of the world. Second, they can be taken to concern not the analysis of the meanings of words but of extralinguistic entities: concepts. Read in this latter way, the claims would be interesting if we were inclined to think that conceptual analysis can reveal synthetic *a priori* truths. But today we are not so inclined.

Alternatively, these claims, which are ostensibly about meanings, can be seen as directed at the epistemology of motion.[3] Or talk about meanings can be seen as an opaque way of addressing the adequacies and inadequacies of theories of motion that use one or another spatiotemporal structure, a not implausible reading of Newton's intentions in "De gravitatione," which uses thought experiments to reveal alleged inadequacies in Descartes's theory of motion (see chapter 4).

2 Arguments from Epistemology

The relationists' litany includes repetition of the fact that knowledge of motion rests on perceptions of the relative changes of positions of bodies, from which they want to conclude that motion can only be relational bodily motion. Leibniz, curiously enough, seems to be something of an exception. In a letter to Huygens, he begins with the standard litany: "Even if there were a thousand bodies, I still hold that the phenomena could not provide us (or angels) with an infallible basis for determining the subject or the degree of motion and that each body could be conceived separately as being at rest" (Loemker 1970, p. 418). But he then seemingly betrays the relationist position by adding that "each body does truly have a certain degree of motion, or if you wish, of force." How Leibniz's doctrine of "force" leads to this seeming betrayal will be discussed in chapter 6.

In any case, the attempted epistemological deduction of the relational character of motion is too quick. Grant the obvious: that scientific theorizing about motion starts from sense perceptions involving impressions of relative changes of positions and that an appraisal of the adequacy of the end product of theorizing must ultimately rest on sense perceptions. It does not follow without further ado that an acceptable theory of motion can employ only concepts of motion open to direct perceptual inspection or else that if this stricture is transgressed, the offending concepts must be given an instrumentalist interpretation. Some relationists, notably Mach, provide the further ado in the form of a positivistic account of scientific theories. Not all seventeenth- and eighteenth-century relationists were so narrow minded. Unlike Mach, Huygens and Leibniz were realists about the motions of particles too little to see with the naked eye or with the aid of a microscope.

This leads one to suspect that underneath the epistemological gloss Huygens and Leibniz sometimes put on their arguments are considerations that stand independently of observability. Consider Huygens's dig at Newton's form of absolute motion: "Those who imagine a true motion without respect to other bodies have realized that motion cannot be discerned or distinguished in those bodies which move uniformly in free motion, since in that infinite space they see as immovable, the senses do not find anything which could give rise to such a judgment" (Huygens 1888–1950, 16 : 226). Although stated in terms of the unobservability of uniform absolute motion, the nub of Huygens's objection is independent of what we or creatures with more acute sensory apparatus can or cannot perceive; rather, the point is that, even on Newton's own terms, absolute velocity is otiose in Newtonian mechanics.[4] However, absolute acceleration in general and absolute rotation in particular are not otiose in Newtonian mechanics, and both Huygens and Leibniz responded to this point with ingenious but ultimately untenable considerations (see chapter 4).

3 Arguments from Scientific Theorizing

When winnowed, the remaining kernel of the arguments from meaning and epistemology comes to this. The relationist asserts that no more structure than is present in Machian or perhaps Leibnizian space-time is needed to support an adequate scientific theory of motion, while the absolutist retorts that at least as much structure as in, say, neo-Newtonian or even full Newtonian space-time is needed for an adequate theory of motion. Framing the issue in this way does not lead to a speedy resolution; on the contrary, at first blush it seems to open a Pandora's box of contentious issues.

The trouble begins with trying to get a fix on the conditions that make a theory of motion adequate. It is uncontroversial that at a minimum such a theory must "save the phenomena." Such differences over what constituted the relevant phenomena for seventeenth-to-nineteenth-century scientists need not detain us, save to remark that before the success of Newton's theory it was not apparent that a single theory could provide a unified treatment of terrestrial and celestial motions. But what it means to *save* the phenomena is quite another matter. The reader innocent of the controversies in the philosophy of science may be astounded to learn that there is still no general agreement on this matter. In *The Scientific Image*, one of the more influential philosophy-of-science texts of the 1980s, van Fraassen offers that a theory saves the phenomena just in case there is a model of the theory inside which all the phenomena fit. But this sense of 'save' seems to amount to no more than accommodation, and a very weak sense of accommodation at that, namely, logical consistency with the phenomena, whereas the sense of 'save' that most natural philosophers have had in mind requires not only passive accommodation but also active prediction, systematization, and explanation. Now the lid is fully off the box. Must scientific prediction follow or at least allow for a deterministic pattern? On the answer to this question turns an interesting argument linking the structure of space-time to the issue of substantivalism (see section 6 below). What is a scientific explanation? Is it equivalent to prediction or retrodiction via laws of nature, or must it also plumb the causal structure of the world?[5] And in any case, what is a law of nature? Is it merely a statement of a Humean regularity, or is it a statement of a contingent but real connection between universals or, alternatively, a statement of a non-Humean physical necessity?[6] Even assuming, *mirabile dictu*, an agreement on all of these matters, there still remains the problem of how to explicate and apply the other commonly touted virtues of theories, such as simplicity, coherence, plausibility, fecundity, computational tractability, etc.

If progress on the absolute-relational controversy had to wait upon a resolution of these interminably debated questions of methodology, then all hope of progress would have to be abandoned. Fortunately, the dialogue on the nature of motion can proceed without first having to settle the most problematic methodological questions, and indeed, this dialogue can help to advance the philosophy of scientific methodology.

Symmetry principles provide the most useful initial guide to assessing the relative adequacies of absolute and relational theories of motion. Two symmetry principles are formulated in section 4, and sections 5 and 6 show how these principles apply to the absolute-relational debate.

4 Symmetry Principles

Since the structure of classical space-time is supposed to remain "similar and immovable," we can speak of *the* symmetries of space-time. The intended models $\mathfrak{M}_T$ of a classical theory of motion have the form $\langle M, A_1, A_2, \ldots, P_1, P_2, \ldots \rangle$, where the absolute objects A_i (see the appendix to chapter 1) are geometric-object fields characterizing the fixed space-time structure and the dynamic objects P_j are geometric-object fields characterizing the physical contents of space-time. Intuitively, the A_is are supposed to be the same in each dynamically possible model, while the P_js are allowed to vary from model to model. Newton apparently held a very literal interpretation of 'sameness'. For him the space-time is given once and for all as an emanative effect of God, and any talk about alternative possible worlds or models must be construed as talk about different arrangements of matter within this fixed space-time. For our purposes it is sufficient that the space-time is the same in every element of $\mathfrak{M}_T$ in the sense that for any $\langle M, A_1, A_2, \ldots, P_1, P_2, \ldots \rangle, \langle M', A'_1, A'_2, \ldots, P'_1, P'_2, \ldots \rangle \in \mathfrak{M}_T$, there is a diffeomorphism d that maps M onto M' in a way that $d * A_i = A'_i$ for all i. Then in explicating the notion of symmetries we can assume without loss of generality that the space-time is literally the same throughout $\mathfrak{M}_T$. A *space-time symmetry* of the fixed space-time is a mapping that leaves all of the A_is invariant, i.e., a diffeomorphism ψ that maps M onto M in a way that $\psi * A_i = A_i$ for all i.

Consider a model $\mathcal{M} = \langle M, A_1, A_2, \ldots, P_1, P_2, \ldots \rangle$ and let Φ be a diffeomorphism that maps M onto M. Define $\mathcal{M}_\Phi \equiv \langle M, A_1, A_2, \ldots, \Phi * P_1, \Phi * P_2, \ldots \rangle$. Now Φ will be said to be a *dynamic symmetry* of T just in case for any $\mathcal{M} \in \mathfrak{M}_T$, it is also the case that $\mathcal{M}_\Phi \in \mathcal{M}_T$. The relation between this model-closure principle and the more usual conception of dynamic symmetry as expressing the equivalence of frames can best be seen in terms of specific examples. Thus, suppose that the space-time is neo-Newtonian space-time (sec. 2.4), and let $\Phi(\mathbf{v})$ be the three parameter family of inertial frame boosts corresponding to the proper Galilean coordinate transformations

$$\mathbf{x}' = \mathbf{x} - \mathbf{v}t, \qquad t' = t. \tag{3.1}$$

The $\Phi(\mathbf{v})$ form a group, with $\Phi(\mathbf{v}_1) \circ \Phi(\mathbf{v}_2) = \Phi(\mathbf{v}_1 + \mathbf{v}_2)$, $\Phi(0) = \text{id}$, and $\Phi^{-1}(\mathbf{v}) = \Phi(-\mathbf{v})$. Choose two inertial frames whose relative velocity is $\mathbf{v}$. Then $\mathcal{M}_{\Phi(\mathbf{v})}$ represents a physical situation that has the same relation to the "moving" frame as that represented by $\mathcal{M}$ has to the "stationary" frame. For example, if $\mathcal{M}$ describes a particle that remains at rest at the origin of the stationary frame, then $\mathcal{M}_{\Phi(\mathbf{v})}$ describes a particle that remains at rest at the origin of the moving frame. That the group of Galilean-frame boosts are the symmetries of T in the above sense means that the T-lawlike behavior of physical systems does not distinguish among the inertial frames, for whatever behavior is allowed by the laws of T relative to one inertial frame is also allowed relative to any other inertial frame.[7]

We are now in a position to formulate two symmetry principles that, it should be emphasized, are not "meaning postulates" on symmetries but are rather conditions of adequacy on theories of motion.

SP1 Any dynamical symmetry of T is a space-time symmetry of T.

SP2 Any space-time symmetry of T is a dynamical symmetry of T.

Behind both principles lies the realization that laws of motion cannot be written on thin air alone but require the support of various space-time structures. The symmetry principles then provide standards for judging when the laws and the space-time structure are appropriate to one another. The motivation for (SP1) derives from combining a particular conception of the main function of laws of motion with an argument that makes use of Occam's razor. Laws of motion, at least in so far as they relate to particles, serve to pick out a class of allowable or dynamically possible trajectories. If (SP1) fails, the same set of trajectories can be picked out by the laws working in the setting of a weaker space-time structure. The theory that fails (SP1) is thus using more space-time structure than is needed to support the laws, and slicing away this superfluous structure serves to restore (SP1). In the next section we shall see some concrete examples of (SP1) at work.

Two arguments can be given in support of (SP2). The first is an argument from general covariance. Let us say that the laws of T are *generally covariant* just in case whenever $\mathcal{M} = \langle M, A_1, A_2, \ldots, P_1, P_2, \ldots \rangle \in \mathfrak{M}_T$, then also $\mathcal{M}^\Phi \in \mathfrak{M}_T$, where $\mathcal{M}^\Phi \equiv \langle M, \Phi * A_1, \Phi * A_2, \ldots, \Phi * P_1, \Phi * P_2, \ldots \rangle$ for any manifold diffeomorphism Φ. Then if Φ is a space-time symmetry, i.e., $\Phi * A_i = A_i$ for all i, it follows from general covariance that Φ is a dynamical symmetry. Laws typically called generally covariant, i.e., partial differential equations written in terms of geometric-object fields so as to have a form that is the same in every coordinate system, are ones that are generally covariant according to the above definition. The argument is completed by adding the premise that laws of motion, whatever their specific content and whether Newtonian, Einsteinian or otherwise, are about an intrinsic reality that is independent of coordinate systems, observers, points of view, etc. The argument has noncircular force only to the extent that the latter premise can be supported by an independent characterization of intrinsic reality and a demonstration that a violation of general covariance involves a nonintrinsic reality. With enough fiddling, most equations of motion can be put in a generally covariant form. But the result of doing so does not guarantee any interesting form of dynamical symmetry; indeed, the fiddling may involve the introduction of a space-time structure that, by (SP1), restricts the dynamical symmetries (see section 5 below).

A second argument for (SP2) derives from trying to imagine how (SP2) could fail. Presumably the theory would have to contain names, regarded as rigid designators, of regions of space-time, and the laws of the theory would say that the lawlike behavior that takes place in region R_1 is different from the lawlike behavior that takes place in R_2, even though $R_2 = \Phi(R_1)$ for some space-time symmetry Φ. But such a difference in lawlike behavior is reason to suppose that R_1 and R_2 differ in some structural property that grounds the difference in behavior. The characterization of this structural property in terms of the addition of new elements to the list of A_is means that Φ is no longer a space-time symmetry and (SP2) is restored. Putting the same point slightly differently, it is hard to see how to reconcile a violation of (SP2) with the widely accepted idea that laws of nature must be universal in the sense that the same laws hold good throughout space-time.[8]

5 Absolute Space

If we take Newton at his word in the Scholium, the space-time setting for the theory of motion and gravitation of *Principia* is supposed to be a full Newtonian space-time (section 2.5) whose symmetries are (New). But the dynamic symmetries of this theory are (Gal). Since (Gal) are wider than (New), we have a clear violation of (SP1). Excising the absolute frame from Newtonian space-time yields a neo-Newtonian space-time whose symmetries are (Gal), and (SP1) is restored. This little homily raises a historical puzzle. That the dynamic symmetries are (Gal) is clearly stated by Newton as corollary 5 to the Axioms: "The motions of bodies included in a given space are the same among themselves, whether that space is at rest, or moves uniformly forwards in a right line without any circular motion." Juxtaposing this corollary with the Scholium to the Definitions shows that Newton in effect rejected (SP1). The discussion in section 1 above and in chapter 4 below helps to explain how Newton was driven to this conundrum. But it doesn't help to explain the fact, recently emphasized by Penrose (1987), that in such places as the 1684 manuscript "De motu coporum in mediis regulariter cedentibus" Newton considered elevating Galilean relativity to the status of a basic principle.

Because Newton's laws of motion and gravitation have (Gal) as their dynamic symmetries, no feature of the lawlike behavior of gravitating bodies can be used to distinguish an absolute frame: in that sense, absolute space is unobservable. But what our symmetry considerations suggest is that this objection of unobservability is more accurately stated as an objection based on Occam's razor. In an attempt to give observability considerations some independent status, we can try to imagine that though absolute space plays no essential role in the laws of motions of bodies, it is nevertheless observable: imagine, if you will, that the space-time tracks of the points of absolute space stand out as thin, red lines on the space-time manifold. But then the justification for introducing absolute space can be supplied by appealing to the contrapositive of (SP1): since the laws of optics are not Galilean-invariant (they cannot be under the posited observability of the thin, red lines), the symmetries of space-time cannot include (Gal). A less fanciful example will be examined shortly.

The logical-positivist tradition prevalent in the philosophy of science during the 1920s and 1930s led to the criticism of absolute space on the grounds that statements of the form "Body *b* is undergoing an absolute change of position" are empirically meaningless, because they are not testable (verifiable, falsifiable, or the like). Later empiricists gave up on the attempt to provide a criterion of testability that would apply to individual sentences, and there was a tendency to embrace a form of holism according to which cognitive significance "can at best be attributed to sentences forming a theoretical system, and perhaps rather to such systems as wholes" (Hempel 1965, p. 117). Unfortunately, such a holism would seem to provide a refuge for the believer in absolute space, since, taken as a whole, Newton's original theory was highly testable.

Perhaps the logical empiricists surrendered too easily to the lure of holism. Glymour's (1980) bootstrapping account of theory testing provides means of distributing praise and blame within a theory. Using this approach, one might try to argue that, say, absolute velocity is not an empirically meaningful concept within Newtonian mechanics, for statements about absolute velocities of particles are not bootstrap-testable relative to the theory, since values of the absolute velocities of bodies are not deducible via principles of the theory from values of relative velocities and other observables. Unfortunately, bootstrap-testability is in general too strong a criterion of meaningfulness relative to a theory. For example, statements about the kinetic energies (relative to the rest frame of the container) of individual gas molecules are not bootstrap-testable relative to the kinetic theory of gases, but such statements would surely not be dismissed as empirically meaningless relative to the theory.

I conclude that if we want to say that absolute change of position is not meaningful within Newton's theory, then there is an aspect of meaningfulness better explicated in terms of symmetry considerations than by any known approach based on testability and the like. Further discussion of meaningfulness and invariance is to be found in the appendix to this chapter.

There is no general argument here to the effect that absolute space is, *ipso facto*, metaphysically absurd; indeed, on the proposed reading of symmetry principles, the acceptability of absolute space reduces to the contingent question of whether the world is such that the empirical adequacy of a theory of motion requires a distinguished inertial frame. Nineteenth-century physics provides cases in which, *prima facie*, the evidence seemed to indicate that a positive answer was called for. On closer inspection, however, none of these cases is unproblematic.

As a first example, consider the Fourier equation of heat conduction:

$$\nabla^2\phi = \frac{\partial\phi}{\partial t}, \tag{3.2}$$

where units have been chosen so that the coefficient of thermometric conductivity is unity. If ϕ transforms like a scalar, (3.2) is evidently not Galilean-covariant. We can write the heat equation in generally covariant form while indicating that the standard form (3.2) holds only in a special inertial frame (say, absolute space) by positing that

$$g^{ij}\phi_{\|i\|j} = A^i\phi_{\|i}, \tag{3.3}$$

where g^{ij} is the space metric and A^i is the unit velocity field of absolute space (see chapter 2). In an inertial coordinate system adapted to absolute space, (3.3) reduces to (3.2). The models of the little theory, whose only law is (3.3), have the form $\langle M, g^{ij}, h_{ij}, \Gamma^i_{jk}, A^i, \phi\rangle$, where h_{ij} is the time metric and Γ^i_{jk} is the affine connection (see section 2.5). The dynamical symmetries do not include (Gal), since the models of the theory are not closed under Galilean-velocity boosts; i.e., $\langle M, g^{ij}, h_{ij}, \Gamma^i_{jk}, A^i, \Phi * \phi\rangle \in \mathfrak{M}_T$ does not follow from the fact that $\langle M, g^{ij}, h_{ij}, \Gamma^i_{jk}, A^i, \phi\rangle \in \mathfrak{M}_T$ whenever Φ is a Galilean-velocity boost.

Though mathematically coherent as an example where absolute space does some real work, (3.3) is not a correct presentation of the physics of heat conduction. For ϕ, recall, is supposed to be the temperature of a material medium, e.g., an iron bar, and the standard form (3.2) of the law of heat conduction is supposed to refer to the rest frame of the medium, not to absolute space. Thus, let W^i be the velocity field of the medium, and for the sake of simplicity suppose that the medium executes inertial motion ($W^i{}_{\|j} = 0$). Then instead of (3.3) we should have

$$g^{ij}\phi_{\|i\|j} = W^i\phi_{\|i} \tag{3.4}$$

Then if W^i is included on the right-hand side of the cut for absolute/dynamical (or space-time/physical) contents, the dynamical symmetries of the modified little theory include (Gal), since $\langle M, g^{ij}, h_{ij}, \Gamma^i_{jk}, A^i, \Phi * \phi, \Phi * W^i\rangle$ is a dynamically possible model whenever $\langle M, g^{ij}, h_{ij}, \Gamma^i_{jk}, A^i, \phi, W^i\rangle$ is, and absolute space is rendered otiose.

Similar comments apply to Maxwell's laws of electromagnetism. They too, regarded purely as equations for the electric **E** and magnetic **B** fields do not possess Galilean symmetry. But again, in the nineteenth-century conception **E** and **B** were interpreted as states of a material medium, the ether, and the use of appropriate variables characterizing that medium can restore Galilean symmetry. Nevertheless, nineteenth-century electromagnetism, especially in Lorentz's version, comes much closer than heat conduction to providing a real historical example where the available evidence supported the use of absolute space. For under Lorentz's influence a dematerialization of the ether took place. Lorentz's ether failed the principle of action and reaction, expected to hold for a material medium (Lorentz's hypothesis of no ether drag meant that the ether acted but did not react), and no mechanical properties were attributed to this ether, save to say that it picks out a special frame as its rest frame. Although Lorentz would not have agreed, there is the clear intimation that Lorentz's ether was fading into pure absolute space and that as a result the electromagnetic field can be regarded as an independent entity rather than as the state of a material medium.[9] However, the resulting theory of classical electromagnetism is not free of internal troubles. It is worth working through the details in order to appreciate how difficult it is to construct an interesting and physically well motivated example where absolute space plays an indispensible role.

In fact, there are two different approaches to classical electromagnetism. On what I shall call the downstairs version, the *downstairs Maxwell tensor*, as defined by

$$^*F_{ij} \equiv \begin{pmatrix} 0 & B_z & -B_y & E_x \\ -B_z & 0 & B_x & E_y \\ B_y & -B_x & 0 & E_z \\ -E_x & -E_y & -E_z & 0 \end{pmatrix}, \tag{3.5}$$

is required to transform like a covariant tensor under the Galilean transformations (3.1). The associated field transformations, relating the **E** and **B** fields as measured in different inertial frames, are

$$\mathbf{E}' = \mathbf{E} + \mathbf{v} \times \mathbf{B}, \qquad \mathbf{B}' = \mathbf{B}. \tag{3.6}$$

Notice that these transformations do not involve absolute velocities and that they make one pair of Maxwell's equations,

$$\nabla \cdot \mathbf{B} = 0 \qquad \nabla \times \mathbf{E} = -\partial \mathbf{B}/\partial t, \tag{3.7}$$

Galilean-covariant. These equations can be recast in generally covariant form as

$$*F_{[ij\|k]} = 0. \tag{3.8}$$

To write the other two Maxwell equations we need to raise indices on $*F_{ij}$. This is accomplished with the aid of $\hat{g}^{ij} \equiv g^{ij} - A^i A^j$ (see section 2.5). In the source-free cases, the second pair of Maxwell equations can then be written in generally covariant form as

$$*F^{ij}{}_{\|j} = 0. \tag{3.9}$$

In an inertial coordinate system adapted to the absolute frame, (3.9) reduces to the familiar

$$\nabla \cdot \mathbf{E} = 0, \qquad \nabla \times \mathbf{B} = \partial \mathbf{E}/\partial t. \tag{3.10}$$

But in an arbitrary inertial coordinate system the form is more complicated, since absolute velocities enter.

Charges and currents can be introduced by positing that $J^i = (\mathbf{j}, \rho)$, where $\mathbf{j}$ is three-vector current and ρ is the charge density, transforms like a contravector under (3.1). This gives as the Galilean transformation law

$$\rho' = \rho, \qquad \mathbf{j}' = \mathbf{j} - \mathbf{v}\rho. \tag{3.11}$$

Equation (3.9) is generalized to

$$*F^{ij}{}_{\|j} = J^i, \tag{3.9$'$}$$

which in an inertial coordinate system adapted to the absolute frame reduces to

$$\nabla \cdot \mathbf{B} = 0, \qquad \nabla \times \mathbf{B} = \partial \mathbf{E}/\partial t + \mathbf{j}. \tag{3.10$'$}$$

The downstairs Lorentz force law for a particle carrying charge q is

$$*F^i = q h^{im} F_{mn} U^n, \tag{3.12}$$

where U^i is the unit four-velocity of the particle. In keeping with (3.11), q is assumed to be an invariant. The downstairs electromagnetic-force vector $*F^i$ is a space vector ($*F^i t_i = 0$), and in an inertial coordinate system $*F^i = (*\mathbf{F}, 0)$, where

$$*\mathbf{F} = q(\mathbf{E} + \mathbf{u} \times \mathbf{B}) \tag{3.13}$$

and $U^i = (\mathbf{u}, 1)$. It is easily checked that (3.13) is consistent with the downstairs field transformations (3.6):

$$*\mathbf{F}' = q(\mathbf{E}' + \mathbf{u}' \times \mathbf{B}')$$
$$= q(\mathbf{E} + \mathbf{v} \times \mathbf{B} + (\mathbf{u} - \mathbf{v}) \times \mathbf{B})$$
$$= *\mathbf{F}$$

Note that absolute velocities do not enter the downstairs version of the Lorentz force.

In the upstairs approach, the "upstairs Maxwell tensor," as defined by

$$\dagger F^{ij} \equiv \begin{pmatrix} 0 & B_z & -B_y & -E_x \\ -B_x & 0 & B_x & -E_y \\ B_y & -B_x & 0 & -E_z \\ E_x & E_y & E_z & 0 \end{pmatrix}, \tag{3.14}$$

is taken to be the basic object. Requiring that $\dagger F^{ij}$ does indeed transform like a contravariant tensor under (3.1) yields the upstairs field transformations

$$\mathbf{E}' = \mathbf{E} \qquad \mathbf{B}' = \mathbf{B} - \mathbf{v} \times \mathbf{E}, \tag{3.15}$$

which make (3.10′) Galilean-covariant. In generally covariant form, these equations become, in the upstairs approach,

$$\dagger F^{ij}{}_{\|j} = J^i. \tag{3.16}$$

The other pair of Maxwell equations, (3.8), are replaced by

$$\dagger F_{[ij|k]} = 0, \tag{3.17}$$

where the indices on $\dagger F^{ij}$ have been lowered by using the inverse of $\hat{g}^{ij}$. The upstairs version of the Lorentz-force law is

$$\dagger F^i = q h^{im} \dagger F_{mn} U^n \tag{3.18}$$

In an inertial coordinate system adapted to the absolute frame, the spatial part $\dagger\mathbf{F}$ of $\dagger F^i$ has the same form as (3.13), but $\mathbf{u}$ is now the absolute spatial velocity of the particle.

Nothing in the logic of the situation dictates which of these two approaches provides the correct version of classical electromagnetism. In favor of the downstairs version is the fact that it was assumed and experimentally verified that Faraday-induction phenomena do not depend upon absolute motions but only, for example, on the relative motion of a magnetic circuit and a conducting circuit, as emphasized by Einstein (1905) in the opening section of his paper "On the Electrodynamics of Moving Bodies." Since these phenomena are governed by equations (3.6), (3.7), and (3.12), one would want them to be free of absolute velocities, as is guaranteed by the downstairs approach. On the other hand, in the standard, classical understanding of the Lorentz-force law the $\mathbf{u}$ in the $\mathbf{u} \times \mathbf{B}$ term in (3.13) was supposed to refer to the absolute velocity of the particle, which favors the upstairs approach. This approach is also favored by the null results of the first-order magnetic-induction experiment of Des Coudres (1889) and the later second-order experiments of Trouton (1902) and Trouton and Noble (1904) performed in an attempt to detect absolute motion, for the relevant laws for these experiments are (3.10′), which are Galilean-covariant under the upstairs field transformations, (3.15). Thus, success does not greet the attempt to produce a version of classical electromagnetism in which absolute space plays an indispensable and coherent role, by imagining that $\mathbf{E}$ and $\mathbf{B}$ came to be recognized as field quantities in their own right and that optical experiments, such as that of Michelson and Morley, confirmed the law of Galilean-velocity addition for light. These imaginings lead to two incompatible versions of electromagnetism, and to choose between them one needs further imaginings to the effect that either the Faraday or the magnetic-induction experiments yielded nonstandard results. At this point one loses contact with historical reality.

As a matter of actual history, Einstein objected strenuously to Lorentz's dualistic explanation of Faraday induction. In the case of a conducting circuit at absolute rest and a moving magnet, a current is produced because according to (3.7) the changing $\mathbf{B}$ field causes an $\mathbf{E}$ field, which propels the electrons in the circuit, whereas when the circuit is in absolute motion and the magnet at rest, there is, according to Lorentz, no $\mathbf{E}$ field, and the electrons are propelled instead by the Lorentz force $\mathbf{u} \times \mathbf{B}$ (where $\mathbf{u}$ is the absolute velocity of the circuit). With the same relative motions in the two cases, the observed currents are, of course, the same. The downstairs approach described above gives a unified explanation of just the kind Einstein wanted. But, as we have seen, it gives a dualistic treatment of magnetic-induction effects, whereas the upstairs approach suffers from just the opposite asymmetry. There is evidence that attempts to reconcile the asymmetries of electromagnetic induction played a not insignificant role in Einstein's discovery of STR (see Earman et al. 1983). Of course, Einstein was not working with the four-dimensional space-time apparatus used here. Had he been, the route to STR might have been easier and quicker, as Trautman (1966) has noted. In either the upstairs or the downstairs approach, $\hat{g}^{ij}$ is needed to raise or lower indices. When experiments indicate that absolute motion cannot be detected, one natural reaction is to retain $\hat{g}^{ij}$ as a metric of signature $(+, +, +, -)$ but to admit that there is no preferred way to split $\hat{g}^{ij}$ into a g^{ij} part and an $A^i A^j$ part. One thus arrives at the Minkowski metric. The difference between the upstairs and the downstairs approaches disappears, and the sought-after Galilean invariance of all four Maxwell equations is realized, albeit in transmuted form, as Lorentz invariance.

To summarize and repeat, absolute space in the sense of a distinguished reference frame is a suspect notion, not because armchair philosophical reflections reveal that it is somehow metaphysically absurd, but because it has no unproblematic instantiations in examples that are physically interesting and that conform even approximately to historical reality.

6 Symmetries, the Structure of Space-Time, and Substantivalism

An argument by Stein (1977) shows that (SP2) can be used to link the issue of the structure of space-time to the issue of space-time substantivalism. Suppose for the sake of simplicity that the possible choices of structure for classical space-time are those listed in chapter 2. Then if we want to allow for the possibility that particle motions are deterministic and if we want to make a substantivalist interpretation of the space-time manifold, it follows that the structure of space-time must be at least as rich as that of neo-Newtonian space-time (section 2.3), in contradiction of the relationist thesis (R1) (section 1.3). Consider any of the classical space-times weaker than neo-Newtonian space-time. For any such space-time the symmetries are wide enough that there is a symmetry map Φ such that $\Phi = \text{id}$ for all $t \leq 0$ but $\Phi \neq \text{id}$ for $t > 0$. By (SP2), Φ is a symmetry of the laws of motion, whatever they happen to be. This means that for any dynamically possible, $\mathcal{M}$, $\mathcal{M}_\Phi$ is also dynamically possible. By choosing Φ and $\mathcal{M}$ appropriately, we can thus produce two dynamically possible models where the world lines of the particles coincide for all $t \leq 0$ but diverge for $t > 0$, a violation of determinism.[10]

The force of the argument turns on how seriously we should take the possibility of determinism, and this in turn depends upon the form of determinism under consideration. Thus, consider the following argument designed to prove the existence of an absolute frame. If we want to allow for the possibility that the value of a scalar field ϕ at one instant determines its future values and if we want to make a substantivalist interpretation of the space-time manifold, it then follows that the structure of space-time must be at least as rich as that of full Newtonian space-time. Neo-Newtonian space-time will not suffice, for we can choose a Galilean transformation Φ such that $\Phi = \text{id}$ for $t = 0$ but $\Phi \neq \text{id}$ for $t > 0$. The transformation Φ preserves the initial data $\phi(x, 0)$, $+\infty < x < -\infty$, but for appropriate choices of $\mathscr{M}$ and Φ the transformation changes the future values of ϕ. But again by (SP2), $\mathscr{M}_\Phi$ is dynamically possible whenever $\mathscr{M}$ is, so we have a violation of the form of determinism in question. (Aside: The reader who has swallowed the logic of this argument should reflect on the following seeming paradox. The Schrödinger equation of elementary quantum mechanics is deterministic in that $\psi(x, 0)$ determines $\psi(x, t)$, $t > 0$. But the space-time setting for this equation is supposedly neo-Newtonian space-time, so the above argument would seem to apply, which undermines determinism. For the resolution, see chapter 11 of my 1986.)

The former argument is more compelling, both because it appeals to a weaker form of determinism (arguably, the weakest form of Laplacian determinism imaginable) and because seventeenth to nineteenth-century relationists and absolutists alike believed in the deterministic character of particle motions.

Relationists then will want to respond to the former argument, and the knee-jerk response is a *modus tollens*: since all motion is the relative motion of bodies, the structure of space-time cannot be as rich as that of neo-Newtonian space-time, and therefore, the substantivalist interpretation of the space-time manifold must be abandoned. In particular, $\mathscr{M}$ and $\mathscr{M}_\Phi$ must be taken not as corresponding to different physical situations but rather to different descriptions of the same situation. As a result, the relationist must trade an active interpretation of symmetry principles, tacitly assumed in the above presentation, for a passive interpretation; dynamic symmetries, so-called by the absolutist, are not read as freedom to reposition, reorient, or boost physical systems in the space-time container but rather as a freedom to describe the same system in many ways. Leibniz had an array of arguments to support this redescriptivist ploy. The examinations of these arguments will be postponed until chapter 6.

Although the introduction of determinism serves to link issues in the absolute-relational controversy, it cuts no philosophical ice at this juncture, for the absolutist and the relationist will simply draw different morals from the decision to maintain the possibility of determinism. The relationist will be pleased by a nice coherency to relationism; namely, the relationist thesis (R2) (section 1.3) is seen to be a necessary condition for the relationist thesis (R1). Similarly, the absolutist will discern a pleasing coherency to absolutism; namely, in order to maintain that space-time is an absolute (or substantival) being, it must be endowed with a structure sufficiently rich to support absolute acceleration. In chapter 9 I shall argue that determinism does cut ice in the setting of theories in which the structure of space-time is not fixed and immutable.

7 Conclusion

In addition to revealing a linkage between the nature of motion and the issue of the ontological status of space and time, symmetry principles also help to establish an initial victory for the relationist on the first issue: motion is not absolute in Newton's original sense of absolute change of position, or at least all of the evidence available from the seventeenth century to the present day indicates that the concept of absolute change of position is not needed in the construction of empirically adequate theories of physics. Indeed, we had a difficult time in finding any physically interesting and historically accurate examples in which an absolute frame of reference would play an essential and unproblematic role.

However, a relational theory of motion cannot survive merely on a victory over absolute space. That victory forces us out of full Newtonian space-time, but the ground between there and Leibnizian space-time is occupied by the absolutist. Whether or not physics must stand on this ground is a question that historically revolved largely around the nature of rotation. The next chapter reviews the incredible contortions absolutists and relationists alike have gone through in attempts to accommodate the phenomena of circular motion.

Appendix: Comments on Symmetry, Invariance, and Dynamics

Dynamic Symmetries

Suppose that of the objects postulated by T, all and only those objects that characterize the space-time structure, are absolute objects (see the appendix to chapter 2). Suppose further that T is well tuned in that it satisfies both (SP1) and (SP2). Then the dynamic symmetries of the theory (as defined above) will be the manifold diffeomorphisms that leave the absolute objects invariant, as required by the definition of dynamic symmetries put forward by Anderson (1967) (see also Friedman 1973, 1983).

However, when (SP1) is violated, as in Newton's original theory, the Anderson definition yields the unwanted result that the dynamic symmetries are (New). On the contrary, that the dynamic symmetries are (Gal), both intuitively and on the explication offered above, helps to pinpoint what is wrong with Newton's original theory.

Next consider a case where some element of the space-time structure is not absolute, e.g., the affine connection in the Cartan version of Newtonian gravitational theory, in which the space and time metrics remain absolute but the affine structure becomes a dynamic element (see section 2.7). Under the Anderson definition, the dynamic symmetries would be enlarged from (Gal) to (Leib) by the passage from the orthodox treatment of Newtonian gravitation to the Cartan treatment. But that is a potentially misleading conclusion, for it suggests that the equivalence of inertial frames achieved in orthodox Newtonian gravitational theory done in neo-Newtonian space-time has been extended to the equivalence of all rigid frames of reference in Cartan space-time. Whatever extension of the relativity principle is achieved in the Cartan theory, it should be carefully distinguished from the recognizable extension that would occur if laws of motion and gravitation were written in Leibnizian or Machian space-time, neither of which uses any affine connection, absolute or dynamical. Such laws would, in a straightforward and unproblematic sense, treat all rigid frames as equivalent. Laws of this kind are considered in chapter 5.

Invariants and Meaningfulness

McKinsey and Suppes (1955) deserve the credit for recognizing the need for an invariance approach to meaningfulness of mechanical quantities. However, the details of their approach differ considerably from the one I favor. To illustrate the differences, I will concentrate on classical particle theories with a very simple model structure, namely, $\langle M, A_1, A_2, \ldots, W, m, Q_1, Q_2, \ldots \rangle$, where as before the A_is are the absolute objects characterizing the space-time structure, W is a set of world lines of constant mass particles, m is the mass function (i.e., for each $w \in W$, $m(w) > 0$ is interpreted as the mass of the particle), and the Q_js are additional quantities. A *matter isomorphism* Φ from $\mathcal{M} = \langle M, A_1, \ldots, W, m, Q_1, \ldots \rangle$ to $\mathcal{M}' = \langle M, A_1, \ldots, W', m', Q'_1, \ldots \rangle$ is a diffeomorphism of M onto itself that maps W one-to-one onto W', with $m'(\Phi(w)) = m(w)$ for all $w \in W$. The type $\boldsymbol{Q}_k$ is said to be an *I-invariant* for T just in case for any elements $\mathcal{M}, \mathcal{M}' \in \mathfrak{M}_T$, if Φ is a space-time symmetry of T and a matter isomorphism of $\mathcal{M}$ to $\mathcal{M}'$, then the values of $\boldsymbol{Q}_k$ in the two models are related by $\Phi * Q_k = Q'_k$. In words, the I-invariants are the mechanical quantities whose corresponding values in two systems related by a symmetry boost are related by the boost transformation. The name 'I-invariant' is justified by the connection with the notion of implicit definability. Say that $\boldsymbol{Q}_k$ is *implicitly definable* in terms of the space-time and matter structure just in case for any $\mathcal{M}, \mathcal{M}' \in \mathfrak{M}_T$, if $W' = W$ and $m' = m$, then $Q'_k = Q_k$. Assuming that (SP1) holds, the I-invariants coincide with the implicitly definable quantities. First, I-invariants are implicitly definable. For consider two models $\mathcal{M}, \mathcal{M}' \in \mathfrak{M}_T$ such that $W' = W$ and $m' = m$. Take $\Phi = \text{id}$. Since id is trivially a space-time symmetry and a matter isomorphism, it must be the case that $\Phi * Q_k = Q'_k = Q_k$. Second, assume that $\boldsymbol{Q}_k$ is not an I-invariant. Then there are $\mathcal{M}$, $\mathcal{M}' \in \mathfrak{M}_T$ and a space-time symmetry and matter isomorphism Φ such that $\Phi * Q_k \neq Q'_k$. By (SP1), for $\mathcal{M}'_{\Phi^{-1}} \in \mathfrak{M}_T$, $\mathcal{M}'_{\Phi^{-1}} = \langle M, A_1, \ldots, (\Phi^{-1}) * W', (\Phi^{-1}) * m, (\Phi^{-1}) * Q'_k, \ldots \rangle$. Thus, $\boldsymbol{Q}_k$ is not implicitly definable, since $\mathcal{M}$ and $\mathcal{M}'_{\Phi^{-1}}$ have the same W and m but different values of $\boldsymbol{Q}_k$.

For a well-tuned theory satisfying (SP1) and (SP2) I propose that a necessary condition for a quantity to be mechanically meaningful for T is that it be an I-invariant for T. This condition is met by absolute momentum, say, in Newton's original theory, but that theory violated (SP1). It is not satisfied by absolute momentum in Newton's theory done in neo-Newtonian space-time.

To compare this conception with the McKinsey-Suppes approach requires a transcription of their definition of invariance into my formalism. If the transcription I propose is inaccurate, I can only apologize. Call an n-ary quantity $\boldsymbol{Q}_k$ an *MS-invariant* for T just in case for any $\mathcal{M}, \mathcal{M}' \in \mathfrak{M}_T$, if Φ is a space-time symmetry for T and a matter isomorphism from $\mathcal{M}$ to $\mathcal{M}'$, then $Q'_k(\Phi(x_1), \ldots, \Phi(x_n)) \approx Q_k(x_1, \ldots, x_n)$ for any $x_1, \ldots, x_n \in M$. The symbol $\approx$ remains to be explained. When $\boldsymbol{Q}_k$ is scalar-valued, $\approx$ is $=$. When $\boldsymbol{Q}_k$ is not scalar-valued but $\Phi(x_i) = x_i$ for all i, $\approx$ is again $=$. But when $\Phi(x_i) \neq x_i$, we have various choices. For example, when $\boldsymbol{Q}_k$ takes as its values spacelike vector fields and T admits an absolute parallelism, we could take $\approx$ to mean that $Q'_k(\Phi(x))$ is parallel to $Q_k(x)$ and that the magnitudes of the two are the same, or we could take $\approx$ to fail unless $Q'_k(\Phi(x)) = Q_k(x)$. MS-invariance implies I-invariance, but not conversely. For example, in neo-Newtonian space-time the four-vector force $ma^i = m[(d^2x^i/dt^2) + \Gamma^i_{jk}(dx^j/dt)(dx^k/dt)]$ of a particle is an I-invariant but not an MS-invariant, though magnitude of force is an MS-invariant. Thus, taking MS-invariance as a necessary condition for mechanical meaningfulness has the awkward consequence that four-vector force is not mechanically meaningful, even though it is definable, indeed explicitly definable, in terms of the space-time and matter structure. This observation can be extended to a general argument against MS-invariance as a necessary condition for mechanical meaningfulness in cases where MS-invariance is stronger than I-invariance. Consider a theory T that satisfies (SP1) and that uses a quantity $\boldsymbol{Q}$ that is an I-invariant but not an MS-invariant. If MS-invariance were a necessary condition for meaningfulness, then T would have the property that the value of the meaningless $\boldsymbol{Q}$ is implicitly and perhaps also explicitly determined by the space-time and matter structure. Such a result provides pressure either to drop MS-invariance as a necessary condition for meaningfulness or else to modify the space-time structure so that MS- and I-invariance coincide. A near coincidence occurs in Machian space-time (section 2.1), and perhaps someone of the MS-invariance persuasion will take this as an argument for the relational conception of motion. However, as will be seen in the following chapters, Machian space-time cannot support an empirically adequate theory of motion.

4 Rotation

Westfall (1971) has rightly remarked that circular motion was a riddle mapped in precise quantitative detail by seventeenth-century savants but never fully solved by them. Judging by the writings of such leading lights as Maxwell and Poincaré, nineteenth- and early twentieth-century scientists were also unable to obtain a satisfactory solution. The reasons for this failure have to do with the fact that rotation was a focal point of the absolute-relational debate, and as a consequence, all of the obscurities and confusions to which that debate was heir were sucked into the vortex of rotation. The purpose of this chapter is to review the pre-relativity-theory responses to Newton's bucket experiment, especially the responses of the relationists and their sympathizers. I begin with a brief review of the relevant parts of the Scholium.

1 Newton's Argument from Rotation

Newton's Scholium on space and time closes with a ringing declaration: "But how we are to obtain the true motions from their causes, effects and apparent differences, and the converse, shall be explained more at large in the following treatise. For to this end it was that I composed it." Earlier in the Scholium Newton states that "True motion is neither generated nor altered, but by some force impressed upon the body moved." If 'true motion' is thus understood as motion under the action of impressed forces, then Newton's declaration is a literally accurate description of the program of the *Principia*. Of course, the rub is that Newton defines 'true motion' as motion relative to absolute space, and such motion can take place without the imposition of forces. But before getting involved in the difficulties of the doctrine of absolute motion, let us explore the structure of the Scholium a little more fully.[1]

After defining absolute space and time, Newton admits that "because the parts of [absolute] space cannot be seen or distinguished from one another by our senses, therefore in their stead we use sensible measures of them." But he urges that "in philosophical disquisitions we ought to abstract from our senses, and consider things in themselves, distinct from what are only sensible measures of them," and he goes on to say that absolute and relative motion can be distinguished from one another "by their properties, causes, and effects." Newton's subsequent discussion of the properties and causes of motion do little to support his doctrine of absolute motion, but the effects of absolute motion are a different matter: "The effects which distinguish absolute from relative motion are, the forces of receding from the axis of circular motion. For there are no such forces in circular motion purely relative, but in a true and absolute circular motion, they are greater or less, according to the quantity of motion." There follow two illustrative experiments, the first of which is an actual experiment involving a rotating bucket of water.[2] Newton claims that the true and absolute circular motion of the water can be measured by the endeavour of the water to recede from the axis of rotation, an endeavour evidenced by the concavity of the surface of the water. The second experiment is a thought experiment involving two globes held together by a cord. Here the claim is that even supposing the globes to be situated in an otherwise empty universe, "where there was nothing external or sensible with respect to which the globes could be compared," we could nevertheless determine the quantity of absolute circular motion by means of the tension in the cord. Newton also claims that the direction of rotation can be determined by observing the increments and decrements in the tension of the cord when forces are impressed on alternate faces of the globes.

The standard reading of this part of the Scholium is that Newton is offering the bucket and globes experiments as part of an argument for absolute space and against a relational conception of motion. In a provocative challenge to this reading, Laymon (1978) proposes instead that the bucket experiment was intended to serve much more modest functions, namely, to score against Descartes and to illustrate how absolute motion, already assumed to exist, can be distinguished from mere relative motion. Laymon's alternative reading has three virtues. First, it calls attention to the structure of the Scholium, which begins by postulating absolute space and time and then proceeds to illustrate the application of these concepts in mechanics. Second, it emphasizes the important point made by Koyré (1965, pp. 53–114) and others that Descartes was the principal target of the bucket experiment. When Newton says that the endeavour of the water to depart from the axis of rotation "does not depend upon any translation of the water in respect of the ambient bodies," there can be no doubt that by 'ambient' he meant the immediately surrounding bodies and that he meant thereby to refute Descartes account of true, or philosophical, motion.[3] Third, Laymon's reading reminds us that prior to writing the Scholium, Newton developed theological and kinematic arguments for absolute space. (Recall the "De gravitatione" doctrines that space is an emanative effect of God, that it is immutable because God is immutable, that a coherent account of motion requires that bodies be assigned a determinate velocity, and that determinate velocity requires that the motion be referred to "some motionless thing such as extension alone or space in so far as it is seen to be truly distinct from bodies.")

Nevertheless, there can be little doubt that Newton took the experiments with a bucket and with globes as part of an overall justification of the doctrine of absolute motion. Even in "De gravitatione" the theological and kinematic arguments are interlarded with dynamic considerations of the sort that reappear in the Scholium. Thus, *before* taxing Descartes with the alleged absurdity that "thence it follows that a moving body has no determinate velocity," Newton notes that it follows from Descartes's doctrines that "God himself could not generate motion in some bodies even though he impelled them with the greatest force." He then proposes a thought experiment:

> As if it would be the same whether, with a tremendous force, He should cause the skies to turn from east to west, or with a small force turn the Earth in the opposite direction. But who will imagine that the parts of the Earth endeavour to recede from its center on account of a force impressed only upon the heavens? Or is it not more agreeable to reason that when a force imparted to the heavens makes them endeavour to recede from the center of revolution thus caused, they are for that reason the sole bodies properly and absolutely moved; and that when a force impressed upon the earth makes its parts endeavour to recede from the center of revolution thus caused, for that reason it is the sole body properly and absolutely moved, although there is the same relative motion of the bodies in both cases. And thus physical and absolute motion is to be defined from other considerations than translation, such translation being designated as merely external. (Hall and Hall 1962, p. 128)

Whatever else thought experiments can or cannot accomplish, they do serve to stimulate intuitions about what is plausible and what is implausible. I take it that part of the function of the above thought experiment from "De gravitatione" and the experiment of the globes from the Scholium is to make us see that a relational account of rotation is highly implausible.

Laymon contends that the experiments with the bucket and the globes were not intended by Newton to prove the eixstence of absolute space, "since this existence is already assumed by their explanation" (Laymon 1978, p. 410). What this view neglects is the possibility that Newton intended the argument for absolute space to be, at least in part, an inference to the best explanation of the mechanical phenomena. Without defending its historical accuracy, I will pursue this interpretation because of its pedagogical advantages.

Thus, let us suppose that Newton was offering the following argument for absolute space.

P1 The best explanation of mechanical phenomena in general (and the bucket experiment in particular) utilizes absolute acceleration (and absolute rotation in particular).

P2 Absolute acceleration in general (and absolute rotation in particular) must be understood as acceleration (and rotation) relative to absolute space.

With the aid of hindsight wisdom we know that (P2) is false, since absolute rotation is available in Maxwellian space-time, and absolute acceleration in general is available in neo-Newtonian space-time, neither of which involves absolute space (see sections 2.3 and 2.4 above). Newton's early critics didn't have the conceptual apparatus to make the point in this fashion, but a number of them clearly felt that (P2) was false, as we will see in the sections below. The rub for the relationist is that he must refute not only (P2) but (P1) as well. What, then, did Newton's critics have to say to (P1)? Unfortunately, no straightfoward answer is forthcoming, for Newton's critics tended to follow him in equating absolute space in the sense of a distinguished state of rest with a substantival container space, and they tended to work within the limited possibility set that allowed only that motion is the relative motion of bodies or else it is motion relative to an immobile space. Thus, they thought that to maintain a nonsubstantival conception of space, they had to maintain a relational account of motion and that by refuting absolute space, they would vouchsafe such an account.

Before turning to the details of the responses of Newton's contemporaries to the bucket experiment, I cannot forbear from mentioning some of the straw-man attacks launched by modern commentators. The first step is to replace (P1) above with

P1′ The only possible explanation of the bucket experiment involves absolute rotation.

The next step is to savage (P1′). Mach is often praised for his attack on (P1′). In *Space and Time* Reichenbach writes:

Newton concludes that the centrifugal force cannot be explained by a relative motion, since a relative motion exists between the pail and the water at the beginning as well as the end.... Mach replies that Newton overlooked the fact that the surrounding masses of the earth and fixed stars have to be taken into consideration. The water rotates not only relative to the pail but also relative to these large masses, which may be considered as a cause of centrifugal force. (Reichenbach 1957, pp. 213–214)

A similar sentiment is found in Ernest Nagel's *Structure of Science*:

Newton's argument was severely criticized by Ernst Mach, who showed that it involved a serious *non sequitur*. Newton noted quite correctly that the variations in the shape of the surface of the water are not connected with the rotation of the water relative to the sides of the *bucket*. But he concluded that the deformations of the surface must be attributed to a rotation relative to *absolute space*. However, this conclusion does not follow from the experimental data and Newton's other assumptions, for there are in fact two alternative ways of interpreting the data: the change in the shape of the water's surface is a consequence either of a rotation relative to absolute space or of a rotation relative to *some system of bodies different from the bucket*. (Nagel 1961, p. 209)

But (P1′) is transparently a straw man. No one who has read "De gravitatione" and the *Principia* can seriously hold that Newton neglected the *possibility* that the effects of the bucket experiment are due to rotation relative to the stars. Newton considered the possibility but rejected it, because he thought it so implausible as not to merit serious consideration.[4] Huygens and Leibniz, whom Reichenbach takes as his philosophical heroes, agreed with Newton in that neither ever seriously entertained this possibility (see sections 2 to 4 below). Moreover, Newton had developed a comprehensive and powerful theory of mechanics on the basis of a non-relational conception of motion. It is simply silly to object to this theory on the grounds that it might be possible to save the phenomena without recourse to absolute motion. Those who want to deny that the success of Newton's theory supports the absolute side of the absolute-relational controversy are obliged to produce the details of a relational theory that does as well as Newton's in terms of explanation and prediction, or else they must fall back on general instrumentalist arguments. Not being instrumentalists, Huygens and Leibniz felt obliged to try to sketch a relational theory of rotation.

2 The Huygens–Leibniz Correspondence

The 1694 correspondence between Huygens and Leibniz is almost as remarkable for what is left unsaid as for what is said. In May of 1694 Huygens expressed his negative reaction to Newton's Scholium: "[I] pay no attention to the arguments and experiments of Mr. Newton [to prove the existence of absolute space] in his *Principia Philosophiae*, for I know he is in the wrong, and I am waiting to see whether he will not retract in the new edition of his book, which David Gregory is to procure" (Huygens 1888–1950, 10:614).[5] Leibniz responded in June with his own condemnation of Newton: "Mr. Newton acknowledges the equivalence of hypotheses in the case of rectilinear motion; but as concerns circular motions he believes that the effort made by circulating bodies to recede from the center or from the axis of circulation reveals their absolute motion. But I have reasons to believe that nothing breaks the general law of equivalence" (10:645–646). And he adds the intriguing remark that "It seems to me, however, Sir, that you yourself were formerly of Mr. Newton's opinion as regards circular motion." Huygens's letter of August 24 confirmed Leibniz's recollection: "As far as absolute and relative motion are concerned, I admired your memoir, in that you remembered that formerly I was of Mr. Newton's opinion as regards circular motion. That is true and it was only two or three years ago that I found the truer one" (10:669–670).

The way in which both men continue to circle the issue without revealing any details of their responses to Newton naturally arouses the suspicion that neither was confident that he had found a wholly satisfactory relational treatment of rotation.[6] The suspicion is to some extent confirmed by the fact that despite the importance of the issue, neither man ever published his response to Newton. In Huygens's case this confirmation is not especially strong both because he was notoriously careful in what he published and because his death in 1695 prevented him from polishing his rather fragmentary manuscripts on absolute and relative motion.[7] But these manuscripts contain a confirmation of a different sort; namely, the same declarations of the relational character of motion and the same objections to Newton are made over and over again in slightly varying forms, almost as if writing them several times could make them true. In Leibniz's case the confirmation is stronger. His response to Newton's bucket is set out in detail in "Dynamica de potentia et legibus naturae corporae" (1690s) and part 2 of "Specimen dynamicum" (1695), neither of which he published. More revealingly, Leibniz did not avail himself of the ample opportunities he had in the correspondence with Clarke to broadcast his response. Had the correspondence not been cut short by his death, Leibniz would probably have been forced to reveal more of his hand, for in his later replies Clarke began to zero in on the problem of rotation.

Huygens's and Leibniz's treatments of rotation are interestingly at odds. The only noteworthy agreement is that neither looked to the stars as an escape route for relationism, a disappointment for commentators who like to look for "precursors" of Mach among Newton's early critics. Each starts on a different tack: Huygens's analysis is based on taking the idea of a rigid body seriously, while Leibniz's analysis is based on denying that there can be rigidity in the commonly understood sense.

3 Huygens's Response

Prima facie, Huygens's change of heart is puzzling. If prior to reading Newton's *Principia* he was "of Mr. Newton's opinion as regards circular motion," why didn't the *Principia* reinforce rather than undermine his opinion? (See Bernstein 1984 for a discussion of this point.) I believe that there is less here than meets the eye, though admittedly no firm conclusion can be drawn, because the texts do not allow us to pinpoint Huygens's pre-*Principia* opinion or to say in exactly what respects his opinion changed. A 1668 text states, "Straight motion is only relative between different bodies, circular motion [is] another thing and has its criterion (κριτήριοη) that straight motion does not possess at all" (Huygens 1888–1950, 16:183). But instead of providing further enlightenment on this distinction, the passage concludes with the escape phrase "but I shall speak about it on another occasion." This much is clear, however. The "criterion" that circular motion is supposed to possess is centrifugal force, which for Huygens was a real force, comparable to the force of gravity.[8] So although Huygens had been of Newton's opinion that there is an important difference between straight and circular motion, his opinion didn't accord with the *Principia* on the status of *vis centrifuga* (centrifugal force). More important, there is no indication in Huygens's early works that he believed that the presence of *vis centrifuga* entailed that the body was in absolute rotation in the sense that it was rotating with respect to an immovable container space. In his manuscripts on absolute and relative motion of the 1690s he does say that "Long ago I thought that in circular motion there exists a criterion for true motion" (16:226). But insofar as the placid Huygens can be said to have had a visceral reaction to anything, it was to Newton's version of true motion. It seems unlikely that all of the numerous objections to absolute space discussed in the 1690s manuscripts failed to suggest themselves to the younger Huygens and much more likely that the younger Huygens, like the mature Leibniz and Berkeley, wanted a conception of true motion as relative motion plus the presence of a criterion, or force, without any thought that such a motion implies the existence of a spatial substratum.

If there is a lingering mystery about the evolution of Huygens's views on true motion, there is none whatsoever about the nature of his objections to Newton's version of true motion. If true motion is motion relative to an unmoved space, then, Huygens wanted to know, relative to what does the unmoved space rest? "But in fact your unmoved space is at rest with respect to what? Indeed the idea of rest does not apply to it. Therefore the notion of that unmoved space is false *qua* unmoved" (21:507). There follows what is for Huygens an untypical bit of ridicule.

> It is thus that many common people have the notion of what is said [to be] upwards and downwards, and this neither with respect to the earth nor to anything else. Whence they once concluded that nobody could live on the other side of the earth, because their heads being turned downwards, they would not be able to adhere to the earth but would necessarily fall off. Such a notion seemed very evident according to their opinion, but nevertheless it is false, because downwards and upwards are relative to the center of the earth. (21:507)

The second part of the manuscript puts the objection in more forceful terms.

> But the idea or name of neither motion nor rest applies to that infinite and void space. In fact those who affirm that it is at rest seem to do so for no other reason than because they observe that it would be absurd to say that it moves, whence they thought that it ought necessarily to be at rest. Whereas they should have thought, rather, that neither motion nor rest in any way pertains to that space. Therefore it is incoherent to say that a body is truly at rest or in motion with respect to mundane space, while this same space can be said neither to be at rest, nor does there exist in it a change of place. (21:507)

As far as I am aware, Huygens is alone among Newton's critics in thinking that absolute space in the sense of a substratum is incoherent because without setting off a regress, there is no way to say whether this substratum is moving or at rest. This is an instance where Huygens's aversion to philosophizing should have served him better.

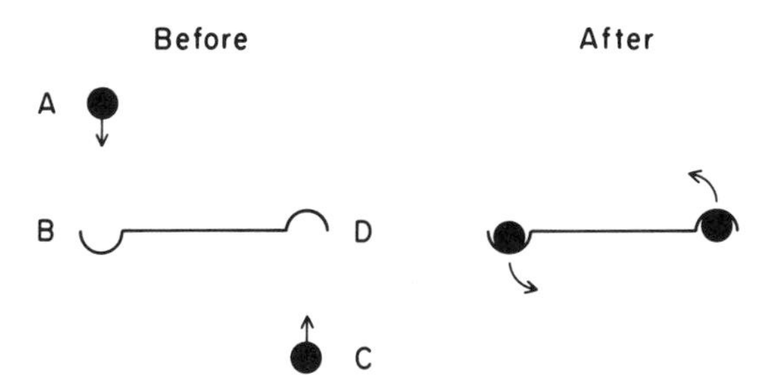

Figure 4.1
Huygens's thought experiment

To his metaphysical objections Huygens adds a second category of objections based on a mixture of epistemology, verificationism, meaning analysis, and conceivability considerations: "To those who ask what motion is, only this answer suggests itself: that bodies can be said to move when their place and distance change, either with respect to each other, or with respect to another body.... We don't understand anything more than this for motion," etc. But since this category has already been explored in chapter 3, I shall move on to Huygens's specific reactions to Newton's argument from rotation.

These reactions can be grouped into a negative and a positive part. This separation is somewhat artificial, since the latter part, containing Huygens's alternative account of rotation, is a direct outgrowth of the former part. The negative part begins by hitting the weak link in Newton's logic:

> It cannot be known how much true motion is contained in each single part [of a rotating body], that is with respect to the space which they conceive as unmoved. For even according to their own judgment, that which moves circularly may at the same time proceed in rectilinear motion either if considered as a whole, or if considered as a conjunction of parts, since [in this last case] the parts are similarly carried around. They themselves grant that rectilinear motion, even if true, cannot be discerned by any sign. (16:227)

Huygens continues to hammer away at this weak link by means of an ingenious thought experiment. In the before picture of figure 4.1 we are to suppose that body *A* moves along the straight line *AB* while body *C* moves along the straight line *CD* parallel to *AB*. "We know that these bodies move with respect to each other and between themselves, but we have to admit nonetheless that we do not know to what extent each of them is truly moved, that is with respect to mundane space" (16:228). Now imagine that a rigid rod with hooks on each end is so placed that body *A* arrives at point *B* at the same moment body *C* arrives at the point *D*. Because the bodies become hooked together by a rigid bond, their original straight-line motion is converted to circular motion. This conversion cannot magically produce true motion where none existed before. "There remains only the relative motion which was before and nothing else occurs except that that which was before in parallel straight lines is now in the opposite parts of the circumference.... Therefore even in such a circular motion nothing else except relative motion can be recognized, as in the motion of the freely moving bodies" (16:229).

This thought experiment leads to Huygens's relational explanation of the nature of circular motion. The circular motion of a body is not to be analyzed as motion relative to some other system of bodies, e.g., the fixed stars, for Huygens makes it clear that his analysis is to apply to a single body, even Newton's rotating globes, presumably. "Circular motion is relative motion along parallel lines, where the direction is continually changed and the distance is kept constant through a bond. Circular motion in one body is the relative motion of the parts, while the distance remains constant owing to the bond" (21:507).

Even those commentators who are not predisposed toward relationism seem compelled to attribute to Huygens some profound insight. Beyond question, Huygens, like many of Newton's critics, saw that the phenomena of rotation do not suffice to ground absolute motion in the sense of absolute change of position. Stein (1977) also wants to attribute to Huygens the "remarkable" insight that there can be an objective notion of velocity difference, or change of direction, even when the distances among the parts of a body remain constant. I am unable to see such an insight clearly emerging from the texts, though I admit that the texts are vague enough that such a reading is possible.[9] Where I really part company with Stein is over his commendation of Huygens for arriving at this insight "not by standing upon philosophical dogmas, whether empiricist or metaphysical ..., but by considering, for a theory known to have fruitful application, exactly how its concepts bear on experience" (Stein 1977, p. 10). But Huygens did not have any theory of dynamics to rival Newton's. And a combination of philosophical dogmas and the tunnel vision that afflicted Newton and his critics alike prevented Huygens from having this remarkable insight in its full and explicit form. For Huygens assumed, as did Newton, that all motion is the relative motion of bodies or else it is motion with respect to an immovable spatial substratum. And as we have seen above, Huygens's philosophical dogmas, both metaphysical and empiricist, seemed to him to militate against the latter alternative and to support the former. Huygens is left with a position that flatly contradicts itself. On one hand, as a good relationist he wants to affirm that "bodies can be said to move when their places and distances change, either with respect to each other, or with respect to another body." On the other hand, he both refuses to refer the motion of a rotating body to some other reference body, and at the same time he has to admit that by the very definition of rigidity the parts of a rigidly rotating body do not change their mutual distances. Further, to analyze rotation in terms of an objective, or absolute, notion of velocity difference rather than objective, or absolute, velocity is to possess exactly the insight Newton lacked, but it is also to reject the full-blown relational conception of motion, something that was beyond the ken of Huygens's philosophical dogmas.

4 Leibniz's Response

If Leibniz had been privy to Huygens's response to Newton, he would have rejected it out of hand. Like Huygens, Leibniz had no inclination to reach for the stars or some other system of external bodies in explaining circular motion. But unlike Huygens, Leibniz recognized that as a consequence the relationist is left in a bind if circular motion is conceived as rigid motion. Indeed, in the "Dynamica" he explicitly concedes that Newton would be correct "if there were anything in the nature of a cord or solidity, and therefore of circular motion as it is commonly conceived." (Gerhardt 1849–1855, 6:508; trans. Stein 1977, p. 42). The only obvious escape route for Leibniz was to deny that there is any true firmness or solidity to bodies. This is not an *ad hoc* move, for Leibniz had left the door open in an earlier essay "Critical Thoughts on the General Part of the Principles of Descartes" (1692). Like Descartes, Leibniz was averse to an explanation of firmness or solidity in terms of a glue that holds the parts of a body together, but at the same time he rejected Descartes's idea that hardness is due to the relative quiescence of the parts of a body and proposed instead that "the primary cause of cohesion is movement, namely, concurrent movement.... It is no doubt also by some kind of magnetism, that is, by an internal coordinated motion, that other parts of certain bodies are linked together" (Loemker 1970, p. 470). And in the "Specimen dynamicum," where he sets out his response to Newton, Leibniz affirms that "firmness is therefore not to be explained except as made by the crowding together by the surrounding matter" (Loemker 1970, p. 449).

Leibniz's second step is to assert that "*all motion is in straight lines, or compounded of straight lines* (p. 449). Putting this assertion together with his analysis of solidity gives his response to Newton.

From these considerations it can be understood why I cannot support some of the philosophical opinions of certain great mathematicians [viz., Newton] on this matter, who admit empty space and seem not to shrink from the theory of attraction but also hold motion to be an absolute thing and claim to prove this from rotation and the centrifugal force arising from it. But since rotation arises only from a composition of rectilinear motions, it follows that if the equipollence of hypotheses is saved in rectilinear motions, however they are assumed, it will also be saved in curvilinear motions. (pp. 449–450)

One possible gloss of Leibniz's argument might go thus. What we see on the macroscopic level as the circular motion of a rigid body is actually the result of the motions of unperceived microscopic bodies that move uniformly and rectilinearly except when they impact upon one another. Moreover, the equipollence of hypotheses is preserved for these micro-motions; in modern jargon, the laws of impact do not assume a distinguished state of rest but are Galilean-invariant.[10] Therefore, the resultant circular macromotion does not require a distinguished state of rest.

This gloss is not quite Leibnizian, since it fits better with a world of atoms moving in the void than with a Leibnizian plenum, but I shall not pause to consider how to make the construction more Leibnizian. For whatever the details of the construction, it is potentially lame, since without further explanation it is not apparent what would occasion the otherwise miraculous coordination of the microbodies needed to produce the perceived macroscopic rigid motion when the body is, say, struck on its circumference. And to the modern reader the most glaring inconsistency in Leibniz's analysis is that it assumes an absolute or invariant notion of straight-line motion, a notion that is simply unavailable in the relationally acceptable Machian and Leibnizian space-times (See sections 2.1 and 2.2 above). It is available in neo-Newtonian space-time (section 2.4), but then so are nonrelational notions of acceleration and rotation.

It is open to Leibniz to respond that he can do without an absolute notion of rectilinear motion, or indeed any absolute quantity of motion. But to make such a response more than a boast, it has to be shown how to save the phenomena by means of laws of motion that are invariant under the Mach or Leibniz transformations. Nontrivial action-at-a-distance laws can be formulated with such invariance properties, but whether or not they can save the phenomena is another matter (see chapter 5). In any case, Leibniz's philosophy prohibits action at a distance, and there is not a hint of how to construct empirically adequate laws of impact using only the weak structures of Machian or Leibnizian space-time.

5 Berkeley's Response

Nearing the end of his life and in poor health (he was recovering from an attack of gout), Leibniz dismissed Berkeley in two vinegar-laced sentences: "The Irishman who attacks the reality of bodies seems neither to offer suitable reasons nor to explain his position sufficiently. I suspect that he belongs to the class of men who want to be known for their paradoxes" (Loemker 1970, p. 609).

Berkeley's immaterialism was generally misunderstood by his contemporaries, and during his lifetime he was never able to shake the disparaging labels of paradoxer and skeptic. In the twentieth century Berkeley has been elevated to the status of Great Man of philosophy, and with this rise in reputation has gone a rise in the estimation of his analysis of motion. It is not uncommon to see Berkeley treated as a precursor (to use Popper's [1953] word) of Mach and Einstein. I will argue that Berkeley's critique of the bucket experiment was every bit as effective as his use of tar water to treat the bloody flux in Ireland.

Sections 114 and 115 of the *Principles* (1710) contain Berkeley's earliest response to Newton's treatment of rotation.

> As to what is said of the centrifugal force, that it does not at all belong to circular relative motion, I do not see how this follows from the experiment which is brought to prove it.... For the water in the vessel at the time wherein it is said to have the greatest relative circular motion, has, I think, no motion at all; as is plain from the foregoing section.... For, to denominate a body "moved" it is requisite, first, that it change its distance or situation with regard to some other body; and secondly, that the force or action occasioning that change be applied to it. (Berkeley 1710, p. 79)

Whether or not Newton violates the "sense of mankind" or the "propriety of language" is not at issue. What is at issue is the correct scientific explanation of circular motion, and Berkeley's *Principles* offers no such explanation. Moreover, the stage of the bucket experiment Berkeley focuses on—the initial stage when the motion of the bucket has not been communicated to the water and the surface of the water is flat—is the least important for Newton's purposes. If this is the best Berkeley had to offer, it is little wonder that Leibniz and other contemporaries were ready to dismiss him as a crank.

Section 116 contains a potentially more interesting but misleading attack on absolute space. "The philosophic consideration of motion does not imply the being of *absolute space*, distinct from that which is perceived by the sense and related to bodies.... And perhaps, if we inquire narrowly, we shall find that we cannot even frame an idea of *pure space* exclusive of all body. This I must confess seems most impossible, as being a most abstract idea" (p. 79). To see what is misleading about the last sentence of the quotation, recall that the prime target of Berkeley's attack on abstract ideas was Locke's *Essay on Human Understanding.* Locke had asked, "Since all things that exist are only particulars, how come we by general terms?" And his answer was, "Words become general by being made the signs of general ideas." Berkeley's response was that "a word becomes general by being made a sign, not of an abstract general idea, but of several particular ideas" (p. 10). As an attack on properties or universals, this does not apply to absolute space, which is not supposed to be a universal but a collection of particulars.

What Berkeley meant to argue, and what comes out more clearly in the later work *De motu* (1721), is that 'absolute space' is meaningless, i.e., denotationless. In the *Principles* Berkeley ties his theory of ideas to a tripartite division of mental activity into sensibility, intellect, and imagination. "It is evident to anyone who takes a survey of the *objects* of human knowledge that they are either ideas actually imprinted on the senses, or else such as are perceived by attending to the passions and operations of the mind, or lastly, ideas formed by help of memory or imagination—either compounding, dividing or barely representing those originally perceived in the aforesaid ways" (p. 22). This apparatus is applied to "absolute space" in section 53 of *De motu* to yield the announced conclusion.

[Absolute space] seems therefore to be mere nothing. The only slight difficulty arising is that it is extended, and extension is a positive quality. But what sort of

extension, I ask, is that ... no part of which can be perceived by sense or pictured in the imagination. For nothing enters the imagination which from the nature of the thing cannot be perceived from sense.... Pure intellect, too, knows nothing of absolute space. That faculty is concerned only with spiritual and inextended things. ... From absolute space then let us take away now the words of the name, and nothing will remain in sense, imagination, or intellect. Nothing else then is denoted by these words than pure privation or negation, i.e., mere nothing. (Berkeley 1721, p. 45)

The same argument, if successful, would show that many of the theoretical terms of modern science are denotationless.

The part of *De motu* most commentators find exciting is section 59.

Then let two globes be conceived to exist and nothing corporeal besides them. Let forces be conceived to be applied in some way; whatever we may understand by the application of forces, a circular motion of the two globes cannot be conceived by the imagination. Then let us suppose that the sky of the fixed stars is created; suddenly from the conception of the approach of the globes to different parts of that sky the motion will be conceived. That is to say that since motion is relative in its own nature, it could not be conceived before the correlated bodies were given. (p. 47)

It is this passage that is supposed to show that Berkeley foreshadowed Mach and Einstein. Whether or not this description is justified will be discussed later. But for now I emphasize that in this section Berkeley is simply illustrating his claim that motion is conceivable only if it is the relative motion of bodies, and he is showing that he has tailored his powers of imagination to fit his philosophical preconceptions. He is not offering a theory of rotation or even claiming that the water sloshes up the sides of the bucket because it is rotating relative to the fixed stars.

One searches *De motu* in vain for a relational account of rotation of the type Huygens and Leibniz struggled to provide. The closest Berkeley comes is in section 69.

As regards circular motion many think that, as motion truly circular increases, the body necessarily tends ever more and more away from the axis. This belief arises from the fact that circular motion can be seen taking its origin, as it were, at every moment from two directions, one along the radius and the other along the tangent, and if in this latter direction only the impetus be increased, then the body in motion will retire from the center.... But if the forces be increased equally in both directions the motion will remain circular though accelerated.... Therefore we must say that the water forced round in the bucket rises to the sides of the vessel, because when new forces are applied in the direction of the tangent of any particle of water, in the same instant new equal centripetal forces are not applied. From which experiment it in no way follows that absolute circular motion is necessarily recognized by the forces of retirement from the axis of motion. (pp. 47–48)

This is the sort of ingenious muddle that those unfortunate enough to have to grade undergraduate papers encounter from the brighter sophomores.

Only in section 65 does Berkeley make a real score against Newton. "The laws of motion ... hold without bringing absolute motion into account. As is plain from this that since according to the principles of those who introduce absolute motion we cannot know by any indication whether the whole frame of things is at rest, or moved uniformly in a direction, clearly we cannot know the absolute motion of any body" (p. 49). But this is the same score that Huygens, Leibniz, and all of Newton's perceptive critics make, and it does not add up to a sufficient defense of relationism.

A reliable yardstick to greatness is the quality of mistakes and failures. Applying this yardstick to the treatment of rotation, Berkeley comes in a very distant fourth to Newton, Huygens, and Leibniz.

6 Kant's Response

The opposition between absolute and relational conceptions of space, time, and motion is a theme that runs like a bright thread through Kant's entire philosophical evolution. In his precritical stage, Kant first accepted and then later rejected a Leibnizian account of space; what turned him against Leibniz was the idea that absolute space was needed to ground the distinction between incongruent counterparts (see chapter 7). But in his critical stage, Kant in turn rejected Newtonian absolute space as an empty figment. The tension generated by juxtaposing the doctrines of the *Critique* with Newton's bucket and the rotating globes permeates the later *Metaphysical Foundations of Natural Science*. I shall address this tension in the present section.

One aspect of the tension is generated by Kant's claim that absolute space is "itself nothing and no object at all" (1786, p. 20) and "nothing but a mere idea" (p. 125), a conclusion that is supposed to follow from the fact that absolute space is not a possible object of experience because "it is not material" (p. 20). In fact, Kant reaches this conclusion by applying a two-pronged test to show that Newton's immovable space cannot be an object of experience; that is, "it cannot be perceived either in itself or in its consequences (motion in absolute space)" (p. 20). At this point in the text

Kant does not alert the reader that the two prongs threaten to yield divergent results. It does seem to follow that, being immaterial, Newton's absolute space cannot be perceived in itself, at least if this means that the parts of this space cannot be detected in the manner of the parts of a rubber ball by reflected light or resistance to touch. But that it can be perceived by its consequences is precisely what Newton's experiments on rotation are supposed to show. The tension of the threatened divergence is soon turned by Kant into a paradox.

Kant's support for a relational account of motion seems to wax and wane and wax again in the *Metaphysical Foundations*. It opens with the uncompromising statement that "all motion that is an object of experience is merely relative" (p. 19). But only a few pages later Kant seems to retreat by applying the relational doctrine to rectilinear, but not to curvilinear, motions: "As concerns nonrectilinear ones, whether I am warranted in regarding a body as moved (e.g., the earth in its daily rotation) and the surrounding space (the starry heavens) as at rest, or the latter as moved and the former at rest, is not in all respects equivalent; this will treated in particular in the sequel" (p. 29). In the sequel, "Metaphysical Foundations of Phenomenology," Kant asserts in proposition 1 that "the circular motion of a matter, in contradistinction to the opposite [rectilinear] motion of the space, is an actual [i.e., objective] predicate of matter" (p. 122). After giving his proof of this proposition, Kant offers an observation on Newton's globes experiment

> On the present subjects, one can refer to the latter part of Newton's Scholium to the definitions with which he begins his *Mathematical Principles of Natural Philosophy*. From this it will become clear that the circular motion of two bodies around a common center . . . even in empty space, and hence without any possible comparison through experience with external space, can nevertheless be cognized by means of experience, and that therefore a motion, which is a change of external relations in space, can be empirically given, although this space itself is not empirically given and is no object of experience. This paradox deserves to be solved. (p. 123)

Indeed it does!

Kant's resolution has two components, a minor and a major one. The minor one offers a sop to absolute space that is also a reference to the *Critique*; namely, absolute space is said to be necessary "as an idea that serves as a rule for considering all motions therein only as relative" (p. 127). The major component starts with a reaffirmation of the relational nature of motion (e.g., p. 128) and then proceeds in a manner that at first seems to bear an uncanny resemblance to Huygens's treatment of the same problem (see section 3 above). For like Huygens, Kant acknowledges that the experiment with the globes induces a distinction between true versus illusive motion, but again like Huygens he refuses to follow Newton in taking this distinction to coincide with that between absolute and relative motion. "This motion [the true motion of the rotating body], even though it is no change of relation in empirical space, is nevertheless no absolute motion but a continuous change of the relations of matters to one another, although it is represented in absolute space and hence is actually only relative motion and, for just this reason alone, is true motion" (p. 129). The passage would not be out of place among Huygens's manuscript notes of the 1690s, at least if we take the phrase "continuous change of the relations of matters to one another" to refer to the relations among the parts of the rotating body. However, in the very same paragraph Kant states that the actual motion is to be "referred to the space enclosed within the moved matter (namely, referred to the center of this matter) but not referred to the external space" (p. 130).

Kant's solution to the riddle of rotation is even lamer than Huygens's and there is no temptation to attribute to Kant some deep insight into the nature of motion. Consistent with Kant's relational conception of motion, the "space enclosed within" cannot be an immaterial Newtonian space. It can only be a material space consisting of—what? Kant does not bother to say, but presumably it must be some sort of subtle matter that, like Lorentz's ether, is not dragged along by ordinary matter. But such subtle matter is nor more the direct object of experience than Newton's absolute space, and thus Kant's solution is unacceptable in its own terms.

7 Maxwell's Response

The fog of the battle fought in the scientific revolution over such fundamental concepts as space and motion was not easily dispersed, and we find it lingering still when we move forward in time a century and a half or more.

James Clerk Maxwell's *Matter and Motion* (1877) was designed, as the preface informs us, as "an introduction to the study of Physical Science in general." As befits such an introduction, it begins with a discussion of the "Nature of Physical Science" (article 1) and proceeds, somewhat tediously, through "Definition of a Material System" (article 2), "Definition of Internal and External" (article 3), "Definition of Configuration" (article 4), etc. But by article 16 Maxwell has moved into deep waters, for here he condemns Descartes's identification of extension and matter and seems to settle on a more Newtonian approach. "We shall find it more conducive to scientific progress to recognize, with Newton, the ideas of time and space as distinct, at least in thought, from that of the material system whose relations these ideas serve to coordinate" (Maxwell 1877, p. 11). Articles 17 and 18 continue the Newtonian line with assertions that are virtual quotations from Newton's Scholium on absolute space and time. This Newtonian reverie is abruptly punctuated by the remark, "All our knowledge, both of time and place, is essentially relative" (p. 12). In an apparent effort to preserve the appearance of consistency, Joseph Larmor has added an explanatory footnote giving a sort of Kantian gloss to Maxwell's intentions; Maxwell's position, he says, "seems to be that our knowledge is relative, but needs definite space and time as a frame for its coherent expression." But Maxwell's intentions at this juncture are far from clear, for he closes article 18 with the remark that "Any one ... who will try to imagine the state of a mind conscious of knowing the absolute position of a point will ever after be content with our relative knowledge" (p. 12).

As has been repeatedly emphasized above, the absolute-relational controversy is in the first instance concerned with the nature of motion. It is thus to Maxwell's treatment of motion that we must turn if we are to discern his true intentions. Article 30 ("Meaning of the Phrase 'At Rest'") seems to put him squarely in the relationist camp. In a relationist pledge of allegiance that could have been lifted directed from Huygens, he asserts, "It is therefore unscientific to distinguish between rest and motion, as between two different states of a body itself, since it is impossible to speak of a body being at rest or in motion except without reference, expressed or implied, to some other body" (p. 22). In article 35 this pledge of allegiance is extended to cover acceleration. "Acceleration," he writes, "like position and velocity, is a relative term and cannot be interpreted absolutely" (p. 25).

The reader now eagerly begins to turn the pages of *Matter and Motion* to find Maxwell's relational treatment of rotation. One is hopeful that what neither Huygens, Leibniz, Berkeley, nor Kant could supply will now be provided as a result of the accumulated wisdom of a century and a half and of Maxwell's unique genius. What the reader actually finds is an abrupt about-face. The crucial paragraph in article 105 on Newton's bucket experiment reads: "The water in the spinning bucket rises up the sides, and is depressed in the middle, showing that in order to make it move in a circle a pressure must be exerted towards the axis. This concavity of the surface depends on the absolute motion of rotation of the water and not on its relative rotation" (p. 85). It would be a mistake, however, to construe this passage as a full endorsement of Newton's absolute space. In the preceding article Maxwell acknowledges that he has tacitly assumed that "in comparing one configuration of the system with another, we are able to draw a line in the final configuration parallel to a line in the original configuration" (p. 83). Maxwell's set of parallel directions is, of course, inertial structure, and in modern terms what he seems to be proposing is that neo-Newtonian space-time is the appropriate arena for the scientific description of motion. This space-time is consistent with his assertion that there is no absolute position or velocity but not with his assertion that there is no absolute acceleration. Consistency eludes Maxwell, just as it eluded his predecessors.

It is worth noting that Maxwell's earlier remarks suggest a position that provides a consistent treatment of rotation and comes closer to reconciling the absolute and relational viewpoints. In modern terms the idea would be to treat motion in the arena of what in section 2.3 I so presciently called Maxwellian space-time, where there is an absolute rotation (which is supported by article 105) but no absolute acceleration in general (which is supported by article 35). Consider what Maxwell says in article 35 in support of his contention that acceleration is relative.

> If every particle in the material universe ... were at a given instant to have its velocity altered by compounding therewith a new velocity, the same in magnitude and direction for every such particle, all the relative motions of bodies with the system would go on in a perfectly continuous manner, and neither astronomers nor physicists using their instruments all the while would be able to find out that anything had happened. (p. 25)

A not implausible reading of this passage is that the laws of particle motion are or ought to be invariant under the Maxwell transformations.[11]

To illustrate how this idea can be implemented in the context of Newtonian action-at-a-distance mechanics, consider a two body system in which the forces obey Newton's third law: $\mathbf{F}_{12} = \mathbf{F}_{21}$, where $\mathbf{F}_{ij}$ denotes the force exerted by the *i*th particle on the *j*th particle. Then Newton's equations of motion,

$$m_1\ddot{\mathbf{r}}_1 = \mathbf{F}_{21} \qquad m_2\ddot{\mathbf{r}}_2 = \mathbf{F}_{12}, \tag{4.1}$$

entail that

$$m_{12}\ddot{\mathbf{r}}_{12} = \mathbf{F}_{21}, \tag{4.2}$$

where $m_{12} \equiv (m_1 m_2)/(m_1 + m_2)$ and $\mathbf{r}_{12} \equiv \mathbf{r}_1 - \mathbf{r}_2$. If we further assume that the forces act along the line joining the particles with a magnitude $f(r)$, $r = |\mathbf{r}_{12}|$, then (4.2) becomes

$$m_{12}\ddot{\mathbf{r}}_{12} = f(r)\hat{\mathbf{r}}_{12}, \tag{4.3}$$

where $\hat{\mathbf{r}}_{12}$ is a unit vector in the direction $\mathbf{r}_{12}$. Equation (4.3) is manifestly invariant under the Maxwell transformations (section 2.4). When the number N of particles is greater than 2, the relative particle equations

$$\ddot{\mathbf{r}}_{ij} = \frac{1}{m_i}\sum_{k \neq i}^{N} \mathbf{F}_{ki} - \frac{1}{m_j}\sum_{k \neq j}^{N} \mathbf{F}_{kj} \tag{4.4}$$

may not admit such a neat presentation as (4.2) and (4.3), but with appropriate restrictions on $\mathbf{F}_{ij}$ they will be invariant under the Maxwell transformations.[12]

Even if one agrees that Maxwellian space-time is the appropriate arena for describing some classical particle interactions of the action-at-a-distance type, relationist thesis (R1) (section 1.3) fails, since this setting presupposes absolute rotation.[13] Nor is it apparent how field laws—in particular, the field laws of electromagnetism that Maxwell himself codified—can be accommodated to Maxwellian space-time.

8 Mach's Response

"Newton's experiment with the rotating vessel of water simply informs us, that the relative rotation of the water with respect to the sides of the vessel produces *no* noticeable centrifugal forces, but that such forces *are* produced by its relative rotation with respect to the mass of the earth and the other celestial bodies" (Mach 1883, p. 284).

No doubt the tendency to see Mach as one of the heroes of natural philosophy inclines some commentators to try to identify precursors of Mach. But if we give up the game of heroes and villains, it becomes a serious question as to whether Mach's predecessors should be regarded as precursors or whether Mach should be regarded as a recapitulator, for in all of Mach's highly touted critique of Newton's argument from rotation, there is very little that is original.

Consider Mach's comment on the Ptolemaic and Copernican systems: "*Relatively* ... the motions of the universe are the same whether we adopt the Ptolemaic or the Copernican mode of view. Both views are, indeed, equally *correct*; only the latter is more simple and more *practical*" (p. 284). This passage could have been lifted directly from Leibniz, who repeatedly claimed that the choice between the Ptolemaic and Copernican hypotheses could only be made on the basis of simplicity and intelligibility.[14]

Mach correctly takes Newton to task for the inconsistency of maintaining Galilean invariance and the existence of absolute space in the sense of a distinguished state of rest. "The Newtonian laws of force are not altered thereby [i.e., by a Galilean transformation]; only the initial positions and initial velocities—the constants of integration—may alter. By this view Newton gave the *exact* meaning of his hypothetical extension of Galileo's law of inertia. We see that the reduction to absolute space was by no means necessary" (pp. 285–286). But again there is nothing here that cannot be found in Huygens, Leibniz, Berkeley, or Maxwell.

Mach also asserts, "No one is competent to predicate things about absolute space and absolute motion; they are pure things in thought, pure mental constructs, that cannot be produced in experience" (p. 280). The thrust of his assertion is contained in Berkeley's condemnation of absolute space as an abstract idea.

And finally, consider such passages as this: "When we say that a body K alters its direction and velocity solely through the influence of another body K', we have asserted a conception that it is impossible to come at unless other bodies A, B, C ... are present with reference to which the motion of the body K has been estimated. In reality, therefore, we are simply cognizant of a relation of the body K to A, B, C" (pp. 281–282). Such epistemological motivations for relationism can be found in the writings of relationists from Huygens on down.

In one respect Mach is less acute than his predecessors; namely, there is no appreciation, such as we have seen in Huygens, Leibniz, Berkeley, Kant, and Maxwell, that rotation poses a special challenge for relationism. Mach saw, perhaps more clearly than Huygens and Leibniz, that relationism requires that the effects of rotation on the water in Newton's bucket be treated in terms of the relative rotation of the water, if not relative to the bucket then relative to the earth or fixed stars. But then with a breathtaking glibness that none of his predecessors, save the Bishop of Cloyne, could emulate, Mach simply asserted that there is no difficulty in producing such

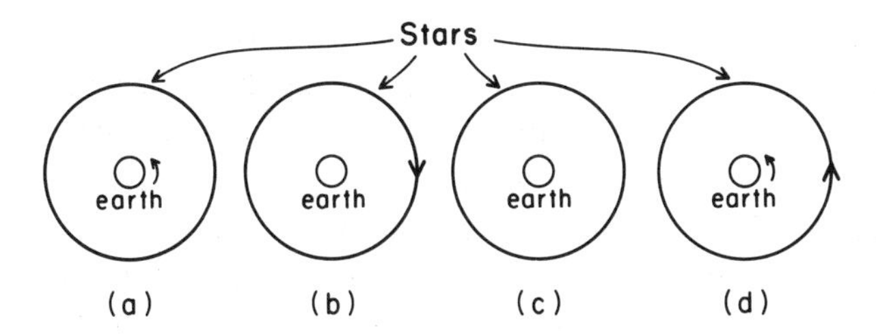

Figure 4.2
Thought experiments for Mach

a relational treatment of rotation: "The principles of mechanics can, indeed, be so conceived, that even for relative rotations centrifugal forces arise" (p. 284). If Mach is asserting that the relationally acceptable Machian or Leibnizian space-times permit the construction of empirically adequate theories of mechanics, one will search *The Science of Mechanics* in vain for the proof of this assertion. Nor is the glibness underwritten by a willingness to take an instrumentalist attitude toward the Newtonian theory of mechanics by accepting the theory, not as a literally true account of the world, but merely as an instrument for predicting effects on relationally well defined quantities, for Newton's theory predicts observationally distinguishable effects for relationally indistinguishable cases, as illustrated by cases (*a*) and (*b*) and again by cases (*c*) and (*d*) in figure 4.2.[15]

Mach balked at imagining the universe to be given twice, much less four times. "The universe is not *twice* given, with an earth at rest and an earth in motion; but only *once*, with its relative motions, alone determinable. It is, accordingly, not permitted us to say how things would be if the earth did not rotate" (p. 284). Mach was perfectly correct, of course, if he meant only to make the obvious point that thought experiments such as Newton's globes rotating in an otherwise empty universe and those illustrated in figure 4.2 cannot be used to settle the absolute-relational dispute. ("When experimenting in thought, it is permissible to modify *unimportant* circumstances in order to bring out new features in a given case; but it is not to be antecedently assumed that the universe is without influence on the phenomena here in question" [p. 341].) But he was mistaken if he meant to suggest that because the only form of directly observable motion is the relative motion of bodies, no observation made in the actual universe, as opposed to the imaginary universe of Newton's experiment with the globes and the thought experiments of figure 4.2, can cut in favor of an absolute over a relational theory of rotation. A Machian theory that makes specific quantitative predictions about how local inertial effects depend upon the distribution of surrounding matter could presumably be tested against standard Newtonian theory either through passive observations on solar and stellar bodies or by experimental protocols whose interventions in nature do not call for anything so radical as altering the rotation of the earth or the heavens. But talk of testing is idle at this stage, since Mach declined to provide the details of such a theory.

Although I have disavowed the game of finding heroes and villains in the history of science, I cannot resist speculating on the game itself by asking why so many commentators want to accord Mach the hero status on the basis of his relatively shallow analysis of the problem of rotation. One understandable but unlaudable reason may be that philosophers, like despots, like to hypnotize the populace with the threats of bogeymen. Mach clearly identified his bogeymen: the "monstrous conceptions of absolute space and absolute time" (p. xxiii). What is not so understandable is how these same commentators swallow the principles that underlie Mach's perceptions of monsters. The same hero who does battle with absolute space and time is also the man who declared, "I can accept the theory of relativity just as little as I can accept the existence of atoms and other such dogma" (p. xxvi).[16] The source of most of Mach's complaints about Newton's doctrines of space and time lies in his perception that Newton failed to be true to his own empiricist tenets. ("Newton has again acted contrary to his expressed intention to investigate *actual facts*" [p. 280].) But the very same narrow empiricism that fuels these complaints also leads to a rejection of atomism, the special and general theories of relativity, and indeed much of twentieth-century physics. Finally, Mach is seen to be a hero because of the reflected glory of Einstein, who is everyone's hero and who supposedly benefited from Mach's precepts in constructing the general theory of relativity. Whatever psychological influence Mach may have had on Einstein, the theory itself does not vindicate Mach's relationism.[17]

9 Poincaré's Response

Chapter 7 ("Relative and Absolute Motion") of Poincaré's *Science and Hypothesis* (1905) begins with an enunciation of a version of the principle of relative motion that seems to come to no more and no less than Galilean relativity, or the equivalence of inertial frames. "The movement of any system whatever ought to obey the same laws, whether it is referred to fixed axes or to the movable axes which are implied in uniform motion in a straight line. This is the principle of relative motion; it is imposed upon us for two reasons: the commonest experience confirms it; the consideration of the contrary hypothesis is singularly repugnant to the mind" (Poincaré 1905, p. 111). Since the Galilean transformations are the symmetries of neo-Newtonian space-time, one would expect Poincaré to endorse all the consequences for the absolute-relational debate that flow from this setting. Instead, we are treated to eleven pages of uncharacteristically tortured discussion.

The stumbling block for Poincaré, rotation, is introduced in a subsection entitled "Newton's argument." Poincaré wonders, "Why is the principle [of relative motion] only true if the motion of the movable axes is uniform and in a straight line? It seems that it should be imposed upon us with the same force if the motion is accelerated, or at any rate if it reduces to a uniform rotation" (p. 113). But the principle fails in these two cases. Poincaré finds no puzzle in the fact that it fails for the case where the motion of the axes is straight but not uniform, for he manages to convince himself that this failure can be explained by the commonplace that the relative motion of two bodies is modified if one or the other is acted upon by an external force.[18] He is considerably exercised, however, by the case of rotation. "If there is no absolute space, can a thing turn without turning with respect to something; and, on the other hand, how can we admit Newton's conclusion and believe in absolute space?" (p. 114). There is a hint here that Poincaré has not entirely escaped the conceptual box that imprisoned Newton and his critics alike: either the motion of a body is judged solely with respect to other bodies or else with respect to absolute space. The suspicion is confirmed by Poincaré's hankering after a version of the principle of relative motion that would legitimate the first alternative. The sought-after principle of relative motion is related to a generalization of the principle of inertia that, in Poincaré's terminology, means that the coordinates of the particle are determined by second-order differential equations, so the generalized principle of relative motion would assert that the differences in the particle coordinates are determined by second-order equations (see pp. 112–113).

There is a seemingly slender but crucial distinction that must be recognized at this point, namely, the difference between quantities such as $\ddot{\mathbf{r}}_{ij}$ and $\ddot{r}_{ij}$. In the preceding section we saw that the empirical content of Newtonian central-force laws of motion can be captured in terms of the quantities $\mathbf{r}_{ij}$, $\dot{\mathbf{r}}_{ij}$, and $\ddot{\mathbf{r}}_{ij}$. But this would not satisfy the epistemologically motivated relationist, for what is directly observed are not these quantities but rather $r_{ij} = |\mathbf{r}_{ij}|$, $\dot{r}_{ij}$, and $\ddot{r}_{ij}$. Or in space-time terminology, the distinction corresponds to that between laws of motion whose proper home is Maxwellian space-time versus those whose proper home is Leibnizian space-time. Rotation militates in favor of the former and against the latter and, thus, against Poincaré's generalized principle of relative motion.

Although Poincaré does not see the point this way, it is clear that these considerations are at the core of his worry. His worry that rotation is inconsistent with the general principle of relative motion is expressed in terms of the following example.

> I assume a system analogous to our solar system, but in which fixed stars foreign to this system cannot be perceived, so that astronomers can only observe the mutual distances of planets and the sun, and not the absolute longitudes of the planets. If we deduce directly from Newton's law the differential equations which define the variation of these distances, these equations will not be of the second order. I mean that if, outside Newton's law, we knew the initial values of these distances and of their derivatives with respect to time—that would not be sufficient to determine the values of these same distances at an ulterior moment.

Poincaré's point can be understood in terms of the two-body problem treated in section 7. Equation (4.3) can be rewritten in terms of $r = |\mathbf{r}_{12}|$ as

$$m_{12}\ddot{r} = f(r) + \frac{L^2}{m_{12}r^3}, \tag{4.5}$$

where L is the magnitude of angular momentum as measured in the center-of-mass frame. This quantity L is what Poincaré dubs an "accidental constant" (p. 118); i.e., it is constant in time but its value is an accidental feature of the system. What worries Poincaré here is that to determine $r(t)$, $t > 0$, we need to know not only $r(0)$ and $\dot{r}(0)$ but also the value of the accidental constant L, or equivalently the values of $r(0)$, $\dot{r}(0)$, and $\ddot{r}(0)$—this is the sense in which (4.5) is not second order.

Poincaré now asks rhetorically why we should hesitate to admit that the subsequent motion depends on the initial value of the second derivative. He answers:

> It can only be because of the mental habits created in us by the constant study of the generalized principle of inertia and of its consequences. The values of the

[relative] distances at any given moment depend upon their initial values, on that of their first derivatives, and something else. What is that *something else*? If we do not want it to be merely one of second derivatives, we have only the choice of hypotheses. (p. 121)

For Poincaré the choice of hypotheses comes to this. We might say that the something else is the "absolute orientation of the universe in space," but that is "not the most satisfactory for the philosopher, because this orientation does not exist" (pp. 121–122). Alternatively, the something else might be Neumann's body Alpha, an equally unpalatable alternative for Poincaré, since "we are destined to never know anything about this body except its name" (p. 122). Poincaré concludes that we should give up this hankering after the something else.

This conclusion not so neatly sweeps the difficulty under the rug. Whether the initial-value problem for the equations of motion is well posed in the usual sense is secondary to the question of whether there are empirically adequate equations of motion that are properly at home in Machian or Leibnizian space-time. Poincaré's remarks do little to help settle this key question.

10 Instrumentalism

A number of late-nineteenth- and early-twentieth-century physicists were gripped by the notion that we need to have an empirical determination of the concept of inertial frame or a way of tying this concept to a concrete system of bodies (see, for example, Lange 1885, 1886, 1902). Apart from the seductive though mistaken idea that legitimate scientific concepts must admit operational definitions, the determination notion seems to be legitimated by the commonsense observation that Newtonian mechanics cannot be applied until we first determine an inertial frame.[19] Like much commonsense wisdom, this bit is misleading. To get started in Newtonian particle mechanics, one needs to know the masses of the particles and the forces acting upon them. But then without first making a determination of an inertial frame, one can proceed to write down Newton's laws of motion for the system and deduce consequences for observables like r_{ij}, $\dot{r}_{ij}$, etc. The equations for these quantities will include additional quantities like L, but at least in the simple two-body case we have seen that these additional quantities can be evaluated in terms of the directly observable relative particle quantities.

These observations emphasize the question, already prompted by Poincaré's discussion, of why we can't dispense with inertial frames and indeed all standards of absolute motion in favor of the observable relative particle quantities, thereby achieving a relational theory of motion. The question is misleading. We can always achieve a relational *description* of motion by milking the original theory for its consequences for relational quantities. The real question is whether the set of such consequences can be structured so as to have the virtues that led one to seek a theory of motion in the first place.

Consider again the central-force problem in Newtonian action-at-a-distance mechanics. Föppl (1897–1910) hypothesized that as measured in the center-of-mass frame, the angular momentum of the universe vanishes. In the two-body problem Föppl's hypothesis has the effect of reducing the Newtonian equation (4.5) to

$$m_{12}\ddot{r} = f(r), \tag{4.6}$$

which contains only good relational quantities. Zanstra (1924) argues that a similar reduction takes place for the general case of $N > 2$ particles. But it is no big surprise to find that the relationist has an easy time of it when troublesome rotation is absent.

Suppose then that L does not vanish. Why can't the relationist take (4.5) as one of his laws of motion? He can, but there are two difficulties in doing so. First, in contrast to the Newtonian, the relationist has no explanation of the fact that the accidental constant L is constant in time. Second, it could be argued that fundamental laws, as opposed to derivative laws, should not contain accidental constants and that, therefore, the relationist has given us the fundamental laws of motion of the system. To meet this second objection, and eventually the first, the relationist can eliminate L by differentiation. The result is to replace a second-order equation (4.5) by the third-order[20]

$$m_{12}(r\dddot{r} + 3\dot{r}\ddot{r}) = \dot{r}(3f(r) + rf'(r)). \tag{4.7}$$

This equation no longer entails that L as originally defined is constant in time, but this need not disturb the relationist, since he is proposing to give up inertial frames and other absolute standards of motion. And the relationist can note that (4.7) does entail that $m_{12}r^3(m_{12}\ddot{r} - f(r))$, which L^2 happens to equal in the two-body case, is constant in time, so that the observational content, if not the original intent, of (4.5) is preserved by (4.7).

All of this the relationist can say, but his story has to change with the value of the particle number N, since the laws of motion for the relational quantities will be different for different values of N. The price for rejecting Poincaré's "something else" is abandoning the unification provided by Newton's theory in the form of a uniform explanation in terms of a single set of laws that apply for all values of N.

This price need not be laid at the doorstep of relationism *per se* but may be seen to result from the instrumentalist approach followed above. Instead of milking Newton's laws for their relational consequences, it is open to the relationist to try to provide a unified treatment of motion by replacing Newton's laws with counterparts that are relationally invariant. Some attempts in this direction will be studied in chapter 5.

11 Conclusion

Newton, Huygens, Leibniz, Berkeley, Maxwell, Kant, Mach, Poincaré—these are names to conjure with. The fact that not one of them was able to provide a coherent theory of the phenomena of classical rotation is at first blush astonishing. It is a testament in part to the difficulty of the problem but in larger part to the strengths of the preconceptions and confusions about the absolute-relational debate. The history of the special and general theories of relativity contains a sequel to this astonishing story of rotation. The sequel is less astonishing but no less important for the absolute-relational debate, or so I shall try to show in the next chapter.

Before closing this chapter, however, it is worth pausing to consider whether it was merely a historical accident that much of the absolute-relational controversy was focused on rotation, for it might be urged, whatever argument or conceptual puzzle can be stated in terms of rotation can be restated in terms of acceleration in a line.[21] My initial response is that if it was a historical accident, it was an accident on such a large scale that it deserves attention for this reason if for no other. The more substantive response is that there are features of rotation that make it an especially difficult challenge for the relationist.

In Newtonian mechanics, acceleration of a body in a line with attendant inertial effects is possible without relative motion of the parts of the body. But according to the theory, such acceleration cannot be achieved without imposing an impressed force, which in turn implies the existence of sources in the form of other bodies. Since the body in question is accelerating relative to other bodies, the way is left open for a relational redescription. By contrast, Newtonian theory says that the inertial effects associated with a rotating body will be experienced in an otherwise empty universe. Rotation, therefore, lends itself better to the kind of thought experiment that, while not settling the issue, served as a powerful intuition pump against relationism.

But let us leave intuition for theory construction. There are ways in which it is harder to achieve an adequate classical theory of rotation as opposed to acceleration in a line. With an absolute or invariant notion of acceleration in a line, rotation becomes absolute, but not vice versa, as Maxwellian space-time illustrates. If we start with Maxwellian space-time and absolute rotation, it is not difficult to treat the motion of bodies moving under central forces without invoking an absolute notion of acceleration in a line. Alternatively, rotation can be suppressed by working in one spatial dimension, in which case it is not hard to work out a viable relational theory of classical gravitational interactions (see section 5.2). But when a fully relational account of rotation is demanded, the setting changes from Maxwellian to Leibnizian or Machian space-time, and in these latter settings an adequate classical theory is more difficult to construct, as we shall see in chapter 5. Chapter 5 also reveals that relativity theory plus a relational account of rotation is an impossible combination.

5 Relational Theories of Motion: A Twentieth-Century Perspective

The history recounted in the preceding chapter might be thought to serve as the basis of an inductive argument against the relationist thesis (R1) (section 1.3); namely, the failure to find a satisfactory relational theory of motion in general and of rotation in particular is evidence that no such theory can be found. Although the general argument form is unappealing, the present instance would seem to carry some probative force, since the failure in question stretches over two and a half centuries and encompasses many of the most brightly shining lights of natural philosophy in those centuries. Nevertheless, the argument has two loopholes. First, none of the exponents of the relational conception of motion realized what is patently obvious in the glare of hindsight wisdom—namely, that Machian and Leibnizian space-times form the natural classical settings for a relational account of motion—and until quite recently there has been no systematic investigation of what possibilities these settings hold. Sections 1 and 2 describe what is now known about these possibilities. Second, even if the relationist dream cannot be fulfilled in classical space-time settings, there remains the hope that relativity theory can be used to vindicate relationism. This hope, often nurtured by nothing more substantial than a confusion between relativity and relationism, will be shown to be a vain one, for the relativistic conception of space-time proves to be much more inhospitable to relationism than the classical conception.

1 Relational Theories of Motion in a Classical Setting

Can there be interesting theories of motion based on classical space-times that do not involve any absolute quantities of motion, whether absolute velocity, acceleration, or rotation? To be interesting, a theory must both be nontrivial and duplicate some of the major explanatory and predictive successes of standard theories of motion.

By way of illustrating the nontriviality requirement, consider *bare space-time*, which is one step down from Machian space-time (section 2.1) in that it contains only absolute simultaneity but no $\mathbb{E}^3$ structure on the instantaneous spaces. No nontrivial theory of motion for a continuous plenum of matter is possible in this setting. A kinematically possible motion for the plenum is specified by a space filling timelike congruence $\mathscr{C}$, a set of smooth timelike curves on the space-time manifold M such that each curve is everywhere oblique to the planes of absolute simultaneity and such that through each point $p \in M$ there passes exactly one curve of $\mathscr{C}$. For any two such congruences $\mathscr{C}$ and $\mathscr{C}'$ there is a symmetry of the bare space-time that carries $\mathscr{C}$ onto $\mathscr{C}'$. Therefore, by symmetry principle (SP2) (section 3.4), either every kinematically possible motion is dynamically possible or none is, i.e., the laws of motion allow that everything goes if anything goes.

The first clear statement of this triviality problem of which I am aware occurs in Weyl's *Philosophy of Mathematics and Natural Science*:

> Without a world structure the concept of relative motion of several bodies has, as the postulate of general relativity shows, no more foundation than the concept of absolute motion of a single body. Let us imagine the four-dimensional world as a mass of plasticine traversed by individual fibers, the world lines of the individual particles. Except for the condition that no two world lines intersect, their pattern may be arbitrarily given. The plasticine can then be continuously deformed so that not only one but all fibers become vertical straight lines. Thus no solution to the problem is possible as long as in adherence to the tendencies of Huyghens and Mach one disregards the structure of the world. (1986, p. 105)

Mistakenly identifying Weyl's "world structure" with inertial structure, Lariviere (1987) contends that Weyl's considerations show that "once we give up the notion of real inertial structure we cannot even talk about relative motion, let alone absolute motion of a single body" (p. 444). The incorrectness of this claim is shown by the fact that the triviality objection does not arise for Machian and Leibnizian space-times, both of which lack inertial structure (see Earman 1989). That the second condition for an interesting theory can be satisfied for Machian and, *a fortiori*, for Leibnizian space-time has been demonstrated only recently.

2 The Relational Theories of Barbour and Bertotti

Barbour (1974, 1982, 1986), Barbour and Bertotti (1977, 1982), and Bertotti and Easthope (1978) have chosen to study the possibilities of a relational theory of gravitation within the setting of Machian space-time. This choice is motivated by a strain of relationism that eschews intrinsic metric structure; in particular, time is seen as a measure of motion. This motivation does not appear to be stable, however, since by symmetry of reasoning the $\mathbb{E}^3$ structure of the instantaneous spaces should also be eschewed, which would force a retreat from Machian to bare space-time, with the unpleasant consequences discussed in section 1. Nevertheless, the choice of Machian space-time is a good one for present purposes, since it offers the relationist the greatest flexibility.[1]

Barbour et al. also choose to investigate action-at-a-distance equations of motion that are based upon an action principle. Barbour (1974) used a Lagrangian of the form $L = VT^{1/2}$, where

$$V = \sum_{i<j}^{N} \frac{m_i m_j}{r_{ij}}, \qquad T = \sum_{i<j}^{N} m_i m_j \dot{r}_{ij}^2, \tag{5.1}$$

and $\dot{r}_{ij} \equiv dr_{ij}/d\lambda$, λ being an arbitrary parameter that increases in the direction of future time. For $N = 1$ there is, of course, no theory of motion, since only relative particle motions are meaningful in this setting. For $N = 2$ all that can be said is that the two particles are approaching or receding from one another, for as a direct consequence of the fact that L is homogeneous to the first degree in $\dot{r}$,[2] the Euler–Lagrange equations are identically satisfied.[3] For $N > 2$ but small, the Euler–Lagrange equations define physically possible motions that are quite different from those defined by Newtonian gravitational theory. But for large N the relational equations of motion can reproduce Newtonian predictions, at least in one spatial dimension. To show this, let $x_i(\lambda)$ be the Cartesian coordinate of particle i. Then $L = V[\sum_{i<j}^{N} m_i m_j(\dot{x}_i^2 - 2\dot{x}_i\dot{x}_j + \dot{x}_j^2)]^{1/2}$. If the time parameter λ is chosen so that $d\lambda = dt \equiv (\sum_{i<j}^{N} m_i m_j dr_{ij}^2)^{1/2}$ (i.e., $T = 1$), and the spatial coordinates are specialized so as to make $\sum_i^N m_i \ddot{x}_i = 0$ (using $d\lambda = dt$), then as Barbour (1974) shows, the Euler–Lagrange equation for the ith particle takes the form

$$\frac{d}{dt}(V m_i \dot{x}_i) = \frac{1}{m}\frac{\partial V}{\partial x_i}, \qquad m \equiv \sum_i^N m_i. \tag{5.2}$$

For a mass-distribution static in the large, V can to good approximation be assumed to be independent of t, in which case (5.2) assumes the Newtonian form

$$m_i \ddot{x}_i = G\frac{\partial V}{\partial x_i}, \tag{5.3}$$

where the gravitational "constant" $G \equiv 1/mV$ depends on the way matter is distributed in the universe. The success of this relational venture is not as impressive as it needs to be, since the restriction to one spatial dimension kills rotation, the *bête noire* of relationism.

In a later article (1977) Barbour and Bertotti modify T to $T' = \sum_{i<j} m_i m_j \dot{r}_{ij}^2/r_{ij}$ so that distant matter exerts less of an inertial influence

than nearby matter. If we choose the time parameter τ (cosmic time) so that $T' = 1$, the Euler–Lagrange equations become

$$V \sum_{j \neq i} \frac{m_j}{r_{ij}} \ddot{r}_{ij} = -\sum_{j \neq i} \frac{m_j}{r_{ij}^2} + \frac{V}{2} \sum_{j \neq i} \frac{m_j}{r_{ij}^2} \dot{r}_{ij}^2 - V \sum_{j \neq i} \frac{m_j}{r_{ij}} \dot{r}_{ij}, \tag{5.4}$$

where the dot or dots now denote differentiation with respect to τ. Ignoring the velocity dependent terms on the right-hand side of (5.4), the analogy with the Newtonian equations of motion indicates that the effective inertial mass of particle i depends upon $V \sum_{j \neq i} m_j / r_{ij}$ and, thus, in keeping with Mach's ideas, upon the distribution of surrounding matter. However, this identification of inertial mass depends upon the time parametrization, and a different choice of parametrization can shift the inertial influence of distant matter into the gravitational "constant" term (see below).

As in Newtonian cosmology, only a finite distribution of matter can be consistently treated, and to square with observation, this distribution must be spherically symmetric. The upshot, as Barbour and Bertotti note, is a dilemma for the relationist: either the spirit of Copernicanism is violated by putting our solar system at the center of the material universe, or else it is put off center, which makes the theory predict nonphysical anisotropic effects. Barbour and Bertotti's response is to derive a local gravitational dynamics in an appropriate cosmological limit. They consider a local system of n particles inside a thin spherical shell of mass M and radius R. Computations are performed in a Cartesian coordinate system in which the shell does not rotate and the origin of which corresponds to the center of the shell. The coordinates of the local masses are denoted by $\mathbf{r}_i$ ($i = 1, 2, \ldots, n$). The cosmological limit is

$$r_i/R \to 0, \qquad m_i/M \to 0, \qquad r_i M / R m_i \to \text{finite}. \tag{5.5}$$

If we take this limit and introduce a local time t related to cosmic time τ by $dt \propto R^2 d\tau$, the local Lagrangian becomes

$$L_{\text{loc}} = \frac{1}{2} \sum_i m_i |\dot{\mathbf{r}}_i|^2 + \frac{4R\dot{R}^2}{M} \sum_{i<j} \frac{m_i m_j}{r_{ij}} + \frac{3R}{2M} \sum_{i<j} \frac{m_i m_j}{r_{ij}} \dot{r}_{ij}^2, \tag{5.6}$$

where the dot now denotes the derivative with respect to t.[4] Equation (5.6) is invariant under the proper Galilean transformations $\mathbf{r} \to \mathbf{r}' = \mathbf{r} + \mathbf{v}t + \mathbf{d}$, and for local processes whose time scale is slow in comparison with the cosmic scale, (5.6) is also approximately invariant under $t \to t' = t +$ constant. To the extent that the third term on the right hand side of (5.6) can be neglected, the equations of motion for local gravitational dynamics are identical in form to the Newtonian equations with a gravitational "constant" of $4R\dot{R}^2/M$.

A somewhat different approach, perhaps even more congenial to Mach's philosophy, is taken by Bertotti and Easthope (1978). In their theory two bodies attract each other via the rest of the universe; specifically, all elementary interactions are three-body interactions, and the interactions between bodies i and j can be thought of as the sum over k of elementary three-body interactions between i, j, and k. The Lagrangian is taken to be

$$L = \left[\sum_{i<j<k} m_i m_j m_k \left(\frac{\dot{r}_{ij}^2 + \dot{r}_{jk}^2 + \dot{r}_{ki}^2}{S_{ijk}} \right) \right]^{1/2}, \tag{5.7}$$

where S_{ijk} is the area of the Euclidean triangle formed by bodies i, j, and k. Taking the cosmological limit and introducing a suitable local time yields a local Lagrangian equation similar to (5.6) except for the addition of two extra velocity-dependent terms.

Are the relational theories of Barbour et al. adequate or at least as adequate as standard Newtonian theory for classical phenomena? 'Classical phenomena' can be taken to include what was or could have been observed by means of nineteenth-century instrumentation. Or more broadly, it could include the phenomena in principle observable that are predicted in some suitable classical limit of GTR. Either way the question does not admit of a neat answer. By suitable adjustments of the parameters R, $\dot{R}$, and M, the Barbour and Bertotti (1977) theory can better standard Newtonian theory by giving the correct values for the orbital period and the perihelion advance of Mercury. But these values in turn have implications for the value of the gravitational "constant," implications that may or may not contradict experiment but are certainly contrary to GTR. The Barbour–Bertotti theory also predicts that, contrary to standard Newtonian theory and GTR, the gravitational action of a spherical body is not the same as if its mass were concentrated at its center, but then standard theory may not be empirically correct. More damningly, the Barbour–Bertotti theory predicts mass-anisotropy effects that contradict standard theory, both classical and relativistic, and experiment.[5]

In addition, there is at present no hint of how to do electromagnetism and nonrelativistic quantum mechanics in Machian space-time, but it is dangerous to prejudge the issue. Indeed, the great merit of the work of

Barbour et al. is that it definitively shows what previous proponents of relationism never came close to showing: that interesting relational theories of classical gravitational phenomena are possible. Such theories may not be adequate to save classical phenomena, but more investigation is needed before that conclusion can be put on a solid foundation.

3 Einstein on Rotation

In "The Foundation of the General Theory of Relativity" (1916), the paper that codified GTR in its final form, Einstein credits Mach with identifying an "epistemological defect" inherent in both Newtonian mechanics and STR. The supposed defect is illustrated by means of a thought experiment that may be taken as a variant on Newton's rotating globes.

> Two fluid bodies of the same size and nature hover in space at so great a distance from each other and from all other masses that only those gravitational forces need be taken into account which arise from the interactions of different parts of the same body. Let the distance between the two bodies be invariable, and in neither of the bodies let there be any relative movements of the parts with respect to one another. But let either mass, as judged by an observer at rest relatively to the other mass, rotate with constant angular velocity about the line joining the masses. Now let us imagine that each of the bodies has been surveyed ..., and let the surface of S_1 prove to be a sphere, and that of S_2 an ellipsoid of revolution. Thereupon we put the question—What is the reason for this difference in the two bodies? No answer can be admitted as epistemologically satisfactory, unless the reason given is an *observable fact of experience.* The law of causality has not the significance of a statement as to the world of experience, except when *observable facts* appear as causes and effects.
>
> Newtonian mechanics does not give a satisfactory answer to this queston. It pronounces as follows:—The laws of mechanics apply to the space R_1, in respect to which the body S_1 is at rest, but not to the space S_2, in respect to which the body S_2 is at rest. But the privileged space R_1 of Galileo ... is a merely *factitious* cause, and not a thing that can be observed. (Einstein 1916, pp. 112–113)

In conformity with Einstein's understanding of the law of causality, the cause of the difference in the shapes of S_1 and S_2 can only lie in the distant masses and the relative motions of S_1 and S_2 with respect to the distant masses.

In one sense, there is nothing original in the quoted passage: Einstein is simply echoing a long relationist tradition with his claim that epistemological considerations demand that the phenomena of rotation be explained purely in terms of the relative motions of bodies. What is more novel is the implicit claim that GTR conforms to this demand. There is no basis for this latter claim, as I trust the following sections will make clear.

4 Rotation and Relativity

A relativistic space-time consists of a manifold M and a Lorentzian metric g on M.[6] In STR, g is an absolute object in that it assumes Minkowskian form in every dynamically possible model. In GTR, g is a dynamic object in that it varies from dynamically possible model to dynamically possible model. This difference will turn out to have important ramifications for the issue of substantivalism (see chapter 9), but for present purposes two similarities between STR and GTR are more pertinent than the differences. First, in contrast to the classical case, where many different affine structures are compatible with the space and time metrics, the relativistic space-time metric g uniquely determines the affine structure through the compatibility requirement: for any g, there exists one and only one affine connection compatible with g. Second, it follows immediately that both STR and GTR ground absolute acceleration and rotation. Consider the world line of a particle, and at any point on that line erect the unit tangent vector V^i ($g_{ij}V^iV^j = -1$). The acceleration vector A^i (not to be confused with the absolute frame of Newtonian space-time) is then defined by $A^i \equiv V^i{}_{\|k}V^k$, where $\|$ denotes covariant differentiation with respect to the uniquely determined affine connection. It is automatic that A^i is a spacelike vector ($A^iV_i = 0$),[7] and its g-norm gives the magnitude of spatial acceleration. Thus, relativity theory in either the special or general form allows one to speak of *the* acceleration of a particle without having to refer explicitly or implicitly to the motion of the particle in question with respect to other particles.

Now consider a congruence of timelike curves that represent, say, the motion of a fluid body, and construct the unit tangent vector field V^i. The associated rotation vector field is defined by $\Omega_i(V) \equiv (1/2)\varepsilon_{ijkl}V^jV^{k\|l}$, where ε_{ijkl} is the unique (up to multiplicative factor) totally antisymmetric covariant tensor of rank four. Here Ω_i is spacelike, since Ω_iV^i is identically zero, and the g-norm Ω of Ω_i gives the magnitude of rotation. Again, we can say whether and how much the body is rotating without making reference to any other body.

At least this is so in the orthodox versions of the theories of relativity. There remains the possibility that there is some nonstandard interpretation of relativity theory that dispenses with absolute quantities of acceleration and rotation. In section 6 below I shall argue that this possibility is foreclosed at least as regards rotation.

The relationist may still wish to claim that GTR indirectly vindicates the relational nature of motion because g is not an absolute object but is determined by the distribution of matter (or matter-energy). This claim will be examined in section 8.

The classical absolutist arguments from rotation rely on the concept of rigid rotation. Rigid motion cannot be relativistically conceived as the rotation of a body that is rigid in the strong sense that it maintains its shape no matter what forces are applied to it, for such a body would transmit signals faster than light. It remains to be discussed how, if at all, the classical concept of rigid motion can be carried over into the relativistic context and how this carryover affects the absolutist's argument from rotation.

5 Relativistic Rigid Motion

Einstein's thought experiment with the two fluid bodies assumed that the bodies move rigidly in that there is no relative movement of the parts of the body with respect to one another. The notion that the distance between the parts of a body remain constant in time makes straightforward sense in the classical space-times discussed in chapter 2, since the distances in question can be measured in the $\mathbb{E}^3$ metric on the planes of absolute simultaneity. The idea of constant spatial distances can be relativistically explicated in several nonequivalent ways, two of which I shall review here.

The most widely accepted definition of relativistic rigid motion is due to Max Born (1909). Consider a congruence of timelike curves represented parametrically as $x^i(\lambda^\alpha, \tau)$, where the λ^α ($\alpha = 1, 2, 3$) label the different curves and τ is proper time along a curve. One computes the spatial distance from one curve of the congruence to an infinitesimally nearby curve as follows. At any point on the chosen curve erect the unit tangent vector $V^i = \partial x^i/\partial\tau$. Project the interval dx^i onto the spacelike hyperplane orthogonal to V^i (see figure 5.1). This can be accomplished by using the projection tensor $k^i{}_j \equiv g^i{}_j + V^iV_j$. Finally, use the space-time metric to calculate the distance ds in the orthogonal hyperplane:

$$ds^2 = g_{mn}(k^m{}_i\, dx^i)(k^n{}_j\, dx^j) \tag{5.8}$$

But $k^i{}_m k^m{}_j = k^i{}_j$, so

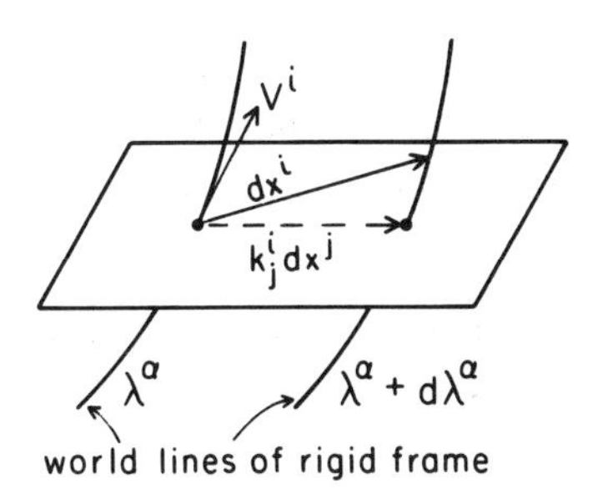

Figure 5.1
Born-rigid motion

$$ds^2 = k_{ij}\,dx^i\,dx^j = k_{ij}\frac{\partial x^i}{\partial \lambda^\alpha}\frac{\partial x^j}{\partial \lambda^\beta}d\lambda^\alpha\,d\lambda^\beta. \tag{5.9}$$

Born rigidity is the requirement that these distances do not change with time, i.e., $d(ds^2)/d\tau = £_V(ds^2) = 0$.[8] But $£_V(d\lambda^\alpha) = 0$, and $£_V(\partial x^i/\partial \lambda^\alpha) = 0$, so that $£_V(ds^2) = £_V(k_{ij})\,dx^i\,dx^j$. Thus, Born rigidity is equivalent to the requirement that $£_V k_{ij} = 0$. A little further calculation shows that this condition is equivalent to $£_{fV} k_{ij} = 0$, where f is a smooth function. Suppose then that the motion is Born-rigid. We can choose a local coordinate system x^1, x^2, x^3, t such that $V^i = (0,0,0,V^4)$. If we take $f = 1/V_4$, $£_{fV} k_{ij} = 0$ is equivalent to $\partial k_{ij}/\partial t = 0$, i.e., the space metric k_{ij} does not change with time.

Another useful characterization of Born rigidity is the vanishing of the expansion $\theta \equiv V^i{}_{\|i}$ and the shear $\sigma_{ij} \equiv k^m{}_i k^n{}_j V_{(m\|n)} - \frac{1}{3}\theta k_{ij}$.[9] Lemma 1 summarizes the various equivalent ways of defining Born rigidity.

LEMMA 1 The following four characterizations of Born rigidity are equivalent:

(i) $d(ds^2)/d\tau = £_V(ds^2) = 0$

(ii) $£_V k_{ij} = 0$

(iii) $£_{fV} k_{ij} = 0$ for any smooth f

(iv) $\theta = 0$ and $\sigma_{ij} = 0$

Another approach to defining relativistic rigid motion uses radar ranging to determine distances. Imagine an observer who remains at rest with respect to some point in the fluid body and who bounces radar signals off the surrounding points. Define the radar distance to another point as one half of the round-trip time, as measured by the proper time of the comoving observer, divided by the speed of light. If the radar distances as measured by all of the comoving observers are constant in time, the congruence will be said to execute *radar-rigid motion.* A useful characterization of radar rigidity was supplied by Müller zum Hagen (1972).

LEMMA 2 A motion is radar rigid if and only if it is stationary, i.e., it generates an isometry of the space-time metric in that $£_{fV}g_{ij} = 0$ for some smooth $f > 0$, or equivalently, $(fV)_{(i\|j)} = 0$.

Putting together lemmas 1 and 2 allows us to derive a general connection between Born and radar rigidity.

LEMMA 3 A motion is radar rigid if and only if it is Born rigid and $A_{[i\|j]} = 0$, where $A^i = V^i{}_{\|j}V^j$ is the acceleration vector.

Lemma 3 holds for any relativistic space-time. In Minkowski space-time there is a more intimate connection between the two concepts of rigidity in the case of rotating motions, as first shown by Herglotz (1910) and Noether (1910).

LEMMA 4 In Minkowski space-time, a rotating motion ($\Omega \neq 0$) is Born rigid if and only if it is radar rigid.

Classically, we can set a body into rotation, making it spin faster and faster, all the while maintaining rigid motion by the use of external forces if necessary. It follows from lemma 4 that this is not possible in Minkowski space-time, at least if we take Born rigidity as our explication of relativistic rigid motion, for in that space-time a body in Born-rigid motion must rotate with a constant angular velocity if it rotates at all.[10] Many early relativists, including Born himself, found this result counterintuitive, and Born (1910) proposed another definition of rigid motion in order to escape it. This Born-again rigidity has disappeared from the literature and will not be discussed here.

In Minkowski space-time a nonrotating Born-rigid motion need not be radar-rigid. The following lemma, due to Malament (1978), gives the general relationship for this case.

LEMMA 5 In Minkowski space-time a nonrotating motion ($\Omega = 0$) is radar-rigid if and only if it is Born-rigid and the magnitude of acceleration is constant.

It can also be shown that

LEMMA 6 In Minkowski space-time a nonrotating motion is Born-rigid if and only if it is orthogonal to a family of flat spacelike hyperplanes.

Together lemmas 4 and 6 give a method for constructing all Born-rigid motions in Minkowski space-time. These lemmas are discussed in the appendix.

The discussion of rigid motions in general-relativistic space-times is more difficult. Indeed, such motions may not exist. An arbitrary general-relativistic space-time may not admit a radar-rigid or stationary motion even locally (i.e., in a finite neighborhood). Pirani and Williams (1963) have studied the integrability conditions for Born-rigid motions in curved space-times and have shown that space-times of Petrov types II, III, and *N* do not admit nonrotating Born-rigid motions. It also follows from their integrability conditions that some curved space-times do not admit any Born-rigid motions, rotating or not, but specific examples are not known. These results confute the classical intuition that by applying external forces if necessary, a small but finite body can be made to execute some form of rigid motion.[11] It also follows that the very geometry of space-time can temporarily thwart the absolutist's pet argument from rotation, namely, that rotation must be absolute rather than relative because the presence of rotation in a body can be detected by observable effects, even though there is no relative motion of the parts of the body if the body is rotating rigidly. This argument is unavailable, of course, where rigid motion is not possible, but the absolutist can repair the damage by noting that the observed effects of rotation cannot always be plausibly explained by the relative motions of the parts of the body (e.g., the strength of the effects of rotation may not correlate with the relative motions).

Other concepts of relativistic rigid motion have been studied by Gardner (1952) and by Synge (1952a, 1952b), who modified Born's definition in an attempt to explain away the positive results of D. C. Miller's repetition of the Michelson–Morley experiment.[12]

6 Relativity, Relationism, and Rotation

Relativity theory, in either its special or general form, is more inimical to a relational conception of motion than is classical physics. The reason traces through the differences between relativistic and classical space-times to that old nemesis for relationism, rotation. We have seen in chapter 2 that there are several classical space-time structures, some of which capture one or another facet of the doctrine that all motion is relative. By contrast, it is difficult to find a space-time structure that is recognizably relativistic and that fails to make the existence of rotation an absolute.

To establish the plausibility of this claim, we must first decide upon a minimal filling for the blank in 'a recognizably relativistic space-time = $M +$ ———'. Here are three candidates for filling the blank: (1) a null cone structure for M; (2) a classification of all tangent vectors of M into three nonoverlapping and mutually exhaustive categories: timelike, spacelike, and null; and (3) a notion of causal connectibility whereby $x, y \in M$ are so connectible just in case there is a smooth causal curve joining x and y.[13] These three choices are not really three, since they are provably equivalent.[14] Also note that any or all of (1) to (3) fixes the space-time metric up to conformal equivalence, where a conformal-equivalence class of Lorentz metrics consists of a maximal class of such metrics such that for any pair g and g' in the class, there is a smooth function $\phi: M \to \mathbb{R}^+$ such that $g' = \phi g$.[15] Finally, I follow Malament (1985) in observing that the existence of rotation is a conformal invariant, i.e., Ω as computed relative to one metric in the conformal class is nonzero just in case it is nonzero relative to all members. This observation can be established by direct computation or more elegantly, as Malament notes, by combining the fact that $\Omega = 0$ just in case the motion is hypersurface-orthogonal with the fact that a conformal transformation preserves orthogonality relations.

The argument does not extend to acceleration in general or to Born-rigid motion. But although its reach is short, it suffices to establish what classical absolutists desperately wanted to prove but never could, namely, that the very idea of space-time in its relativistic guise is irreconcilable with a full-blown relational conception of motion.

7 Einstein's Critique Revisited

It is clear from the foregoing that if Newtonian mechanics and STR are unsatisfactory because they employ the "factitious cause" of inertial frames, then GTR is equally unsatisfactory. Nevertheless, GTR would come closer to fulfilling Einstein's demand that only "*observable facts* ultimately appear as causes and effects" if it sanctioned the use of the fixed stars or, more generally, the frame in which matter is on the average at rest as an inertial frame in that it exhibits the two following properties:

I1 A test particle placed at rest in the matter frame remains at rest.

I2 A test particle given an initial velocity with respect to the frame moves in a straight line with respect to the frame.

To investigate the conditions under which (I1) and (I2) hold in GTR, let us suppose that some suitable averaging process is applied to produce a macroscopic energy-momentum tensor T^{ij} (electromagnetism being omitted from consideration) and the macroscopic velocity field V^i of matter. Suppose first that the average matter distribution behaves as dust, i.e., $T^{ij} = \rho V^i V^j$, where ρ is the density of matter. It then follows from Einstein's field equations for GTR that (I1) holds for the rest frame of matter. The field equations entail the conservation law $T^{ij}{}_{\|j} = 0$, which in the case under consideration implies that

$$\dot{\rho} V^i + \rho\theta V^i + \rho A^i = 0 \qquad (\dot{\rho} \equiv d\rho/d\tau = \rho_{|i} V^i) \tag{5.10}$$

Equation (5.10) says that the acceleration A^i is proportional to the four-velocity V^i, which is only possible if $A^i = 0$, since $A^i V_i = 0$. Property (I1) then follows from the GTR postulate that test particles follow geodesics.

If the averaged-out matter exerts a pressure p, then the macroscopic energy-momentum tensor assumes the more complicated form $T^{ij} = (\rho + p) V^i V^j + p g^{ij}$. In general, the frame comoving with matter will not be geodesic in this setting, and (I1) will fail.

Let us assume that the matter frame is geodesic so that (I1) is secured and proceed to investigate (I2). To make sense of the notion of a straight line with respect to the frame, we need to start with a well-defined notion of the spatial geometry of the space associated with the frame. A point of the associated space is an equivalence class of space-time points, with two points being equivalent just in case they lie on the same curve of the congruence that defines the frame. If the frame is nonrotating and therefore hypersurface-orthogonal, talk about the geometry of the associated space can be given a cash value in terms of the geometry of the orthogonal spacelike hypersurfaces inherited from the space-time metric. If, however, the frame is rotating, this explication is unavailable. But if the frame is Born-rigid, then the space metric k_{ij} can be used to calculate well-defined spatial distances between the points of the space (since $\partial k_{ij}/\partial t = 0$). So let us suppose that the matter frame is Born-rigid. It follows from lemma 3 that the frame is stationary. Our question regarding (I2) now becomes: Does GTR entail that if the matter frame is stationary as well as geodesic, a test particle given an initial velocity relative to the frame will move along a geodesic of the spatial geometry of the associated space? A negative example is provided by the Gödel cosmological model, which is a dust-filled universe satisfying Einstein's field equations, with the rest frame of matter being geodesic, stationary, and everywhere rotating. In a coordinate system comoving with the dust, the Gödel line element takes the form

$$ds^2 = (dx^1)^2 + \tfrac{1}{2}\exp(ax^1)(dx^2)^2 + (dx^3)^2 + 2\exp(2ax^1)\, dx^2\, dt - dt^2, \qquad (5.11)$$

where a is a constant. Evidently, x^3 is an axial coordinate, x^1 is a radial coordinate, and x^2 is an angular coordinate.[16] Now $x^1 = vt$, $x^2 =$ constant, and $x^3 =$ constant is a geodesic of the spatial geometry of the matter frame. But a particle given an initial velocity in the radial direction will not continue to travel in that direction but will spiral outward.[17]

The Gödel model is less than satisfactory as a counterexample because of its pathological causal properties, e.g., every point in the space-time can be connected to itself and to every other point by a future-directed timelike curve, which thus raises the hope that causally well behaved cosmological models in GTR will give a positive answer to our query about (I2). The models of Ozsvath and Schucking (1962) effectively dash this hope.[18]

Another way to bring out the collision between Einstein's epistemological constraint and GTR is to note that the constraint would seem to require that either GTR should not admit solutions in which there are just two bodies in a universe free of matter otherwise, or the class of such solutions shouldn't display asymmetries in the form of a subclass in which both bodies are spherical, another subclass in which one body is more ellipsoidal than the other, and another subclass in which the bodies are equally ellipsoidal. It is difficult to make any firm pronouncements on the topic, since no exact two-body solutions to Einstein's field equations are known, but no general relativist doubts that there are such solutions or that the class of such solutions will display the asymmetries that immediately activate Einstein's epistemological objection. The history of philosophy is littered with failed attempts to use epistemological considerations to limit ontology. Epistemology must learn to live with ontological reality, which often turns out to lie beyond the ken of doctrinaire theories of knowledge.

How could Einstein have been led to try to motivate GTR by means of principles so fundamentally at odds with the completed theory? No doubt part of the explanation lies in the fact that many decades of work lay ahead before a clear understanding of the implications and foundations of the theory could emerge. But a clue to a deeper explanation lies in the title Einstein gave to the section in which the supposed epistemological defect of Newtonian mechanics and STR is discussed: "The Need for an Extension of the Postulate of Relativity." The epistemological objection and Einstein's thought experiment are supposed to motivate a generalization of the special principle of relativity. "Of all imaginable spaces R_1, R_2, etc., in any kind of motion relatively to one another, there is none which we may look upon as privileged *a priori* without reviving the above-mentioned epistemological objection. *The laws of physics must be of such a nature that they apply to systems of reference in any kind of motion*" (Einstein 1916, p. 113). Behind this attempted extension lies the story of a triple confusion. First, in searching for the field equations of GTR, Einstein was guided by the idea that the distribution of matter should determine the metric; the difficulties inherent in this idea will be discussed below in section 8. Second, for a period extending from 1913 into 1915, Einstein convinced himself that matter couldn't determine the metric through generally covariant field equations,[19] and in response he proposed to abandon the requirement of general covariance. After he realized his mistake and reembraced general covariance, he was led to the brink of another confusion: general covariance, he thought, entailed a generalized principle of relativity by which all frames of reference are equivalent; hence the title of section 2 of his 1916 paper and the appellation of "general theory of relativity" for his new theory of gravitation.

None of this detracts from Einstein's monumental achievement, but it should serve as a cautionary tale for philosophers of science who seek to draw wisdom from the philosophical pronouncements of scientists, even the greatest scientists.

8 Mach's Principle

Within the general-relativity industry there is a minor but persistent sub-industry concerned with Mach's Principle. A major object of this sub-industry seems to be to find within GTR some effect that can somehow be associated with something Mach said or could have said. I decry the notion that a principle must contain some element of truth or at least be worthy of serious consideration because the vicissitudes of history have led to a capital ‘p’ in ‘Mach’s Principle’. Nevertheless, the industry surrounding Mach’s Principle has made a significant contribution to our understanding of the foundations of GTR. Since there are excellent review articles on this topic (see, for example, Raine 1981), I do not propose to pursue it, save to briefly discuss a few items directly relevant to the issue of absolute versus relational accounts of motion.

The fact that GTR treats the structure of space-time as dynamic rather than absolute is thought to be attractive to the relationist. Weyl, for example, read Leibniz as having “emphatically stressed the dynamical character of inertia as a tendency to resist deflecting forces” (1966, p. 105). Following Weyl’s lead, Lariviere (1987) proposes to interpret Leibniz not as claiming that there is no real inertial structure to space-time but rather as holding that the inertial structure is dynamic as opposed to absolute. I doubt that the texts will support this reading of Leibniz.[20] But nevertheless, the idea that inertial structure is dynamic is one which modern relationists would find congenial, because it seems to open the way for a relational treatment of relativistic inertial structure through the chain: the distribution of mass determines the space-time metric, which in turn determines the inertial structure, since as seen above, there is one and only one affine connection compatible with a given Lorentzian metric. Despite its seeming appeal, the idea of this chain is hard to instantiate in a form that is both precise and supportive of relationism. In relativity theory it is not mass that matters but mass-energy, and what enters in the field equations of GTR is not mass-energy but energy-momentum. But the idea that energy-momentum determines the space-time metric is no sooner stated than it undermines itself. Suppose for the sake of simplicity that the energy-momentum tensor arises solely from a distribution of dust. That supposition fixes the form of T^{ij} to $\rho V^i V^j$. But this form does not tell us how much energy-momentum is present until it is coupled with the space-time geometry, i.e., $\hat{T}^{ij} = \hat{\rho}\hat{V}^i\hat{V}^j$ represents more energy-momentum than $\tilde{T}^{ij} = \tilde{\rho}\tilde{V}^i\tilde{V}^j$ if and only if $\hat{\rho} > \tilde{\rho}$, at least this is so if $\hat{V}^i$ and $\tilde{V}^j$ are both normed to 1, an operation that requires the use of the space-time metric. This difficulty is avoided in the case of “empty” space, i.e., $T^{ij} = 0$. But the implementation of Mach’s Principle in this case would seem to require that the field equations of GTR admit no solutions for empty space, or alternatively that the solution is unique. In fact, however, Einstein’s field equations, with or without a cosmological constant, admit multiple solutions for $T^{ij} = 0$.

One could declare that vacuum solutions are to be ignored as “unphysical” (see Harré 1986). But if such a declaration is to be consistent with a continued belief in GTR, it should be accompanied by two demonstrations: the first would show that Einstein's field equations can be so modified as to preserve the predictions for the case of $T^{ij} \neq 0$ but to yield no solution or else some standard flat space-time for $T^{ij} = 0$; the second would show that in some appropriate limit in which the gravitational sources are turned off, solutions to Einstein's field equations either become degenerate or else reduce to some standard flat space-time. The second seems unlikely when gravitational radiation is taken into account, and three-quarters of a century of research on GTR has given no support to the first.

The idea that the distribution of mass-energy determines the metric fares even worse in the initial-value problem for Einstein's field equations. If a unique solution is to be determined,[21] the specification of the initial data requires not only a specification of the initial mass-energy distribution but also a specification of the intrinsic spatial geometry of the initial-value hypersurface and its extrinsic curvature, which determines how the surface is to be imbedded into space-time. The former specification is entangled with the latter in two ways. First, as already noted, the very notion of the amount of mass-energy present presupposes metric concepts. And second, Einstein's field equations constrain the initial data through elliptic partial-differential equations that imply that the initial matter distribution cannot be specified independently of the intrinsic spatial geometry and the extrinsic curvature of the initial-value hypersurface.

In *The Meaning of Relativity* Einstein gave a boost to the industry surrounding Mach's Principle by stating that "the [general] theory of relativity makes it appear probable that Mach was on the right road in his thought that inertia depends upon a mutual action of matter" (1955, p. 100). He went on to claim that GTR affirms the expectation that the "inertia of a body must increase when ponderable masses are piled up in its neighborhood." However, the effect that Einstein used to illustrate this idea is nonintrinsic and depends upon the use of a special coordinate system. A second effect mentioned by Einstein is indeed realized in GTR; namely, a rotating mass shell generates within itself Coriolis and centrifugal fields, which gives a prophetic ring to Mach's remark, "No one is competent to say how the [rotating-bucket] experiment would turn out if the sides of the vessel increased in thickness and mass till they were ultimately several leagues thick" (Mach 1883, pp. 284–285). For two separate reasons, however, this effect does not vindicate relationism. First, as already noted, the rotation of the mass shell is an absolute rotation. Second, relationism requires that matter should completely determine inertia and not merely influence it. In the case in point the determination of the local inertial frames inside the rotating mass shell is the result of a dragging along of inertial frames of the background metric, usually assumed to be Minkowskian at spatial infinity. No matter how hard they fiddle, relationists cannot make GTR dance to their tune.

9 Conclusion

Opponents on opposite sides of the classical absolute-relational debate labored mightily to prove that the others' position involved a conceptual incoherency. As regards the nature of motion, they all labored in vain, for the setting of classical space-time is flexible enough to accommodate coherent versions of both views: that all motion is relative motion and that motion involves some absolute quantities, whether velocity, acceleration, or rotation. Empirical adequacy favors the latter view. I choose the word 'favors' advisedly, for no knock-down demonstration is to be found for the conclusion that the best classical theory of motion must use absolute quantities of motion. But the long history of classical physics, coupled with recent investigations of the possibilities afforded by Machian and Leibnizian space-time and our best current understanding of what constitutes a best theory, do lend support to that conclusion.

The advent of relativity theory changes the dimensions of the debate, and it changes them in a manner opposed to the propaganda of the relationist proselytizers who would have us conflate relationism with relativity. For not only do the orthodox versions of STR and GTR fail to vindicate the relational conception of motion; the relativistic space-time setting seems to be unable coherently to accommodate the view that all motion is relative because of its more intimate intertwining of space and time.

Appendix: Rigid Motion in Relativistic Space-Times

Lemma 1

To see that conditions (ii) and (iii) are equivalent, write out the Lie derivative as follows:

$$
\begin{aligned}
\pounds_{fV} k_{ij} &= f k_{ij\|l} V^l + k_{mj}(f V^m)_{\|i} + k_{in}(f V^n)_{\|j} \\
&= f \pounds_V k_{ij} + V^m k_{mj} f_{|i} + V^n k_{in} f_{|j}
\end{aligned}
$$

The last two terms are zero by definition of k_{ij}, which establishes that $\pounds_{fV} k_{ij} = f \pounds_V k_{ij}$.

To see that (ii) is equivalent to (iv), decompose $V_{i\|j}$ as $\sigma_{ij} + \omega_{ij} + \frac{1}{3}\theta k_{ij} - A_i V_j$, where the rotation tensor ω_{ij} is defined by $k^m{}_i k^n{}_j V_{[m\|n]}$. Since ω_{ij} is antisymmetric, condition (iv) is equivalent to

$$V_{(i\|j)} = -A_{(i} V_{j)}.$$

Now compute

$$\begin{aligned}\pounds_V k_{ij} &= k_{ij\|l} V^l + k_{mj} V^m{}_{\|i} + k_{in} V^n{}_{\|j} \\ &= g_{ij\|l} V^l + V_{i\|l} V^l V_j + V_i V_{j\|l} V^l + g_{mj} V^m{}_{\|i} + V_m V^m{}_{\|i} V_j + g_{in} V^n{}_{\|j} \\ &\quad + V_{in} V^n{}_{\|j}\end{aligned}$$

With the identities $g_{ij\|l} = 0$ and $V_i V^i{}_{\|j} = 0$, criterion (ii) reduces to $2A_{(i} V_{j)} + 2V_{(i\|j)} = 0$, which is equivalent to (iv).

Lemma 2

The proof that stationarity implies radar rigidity is straightforward. The converse implication is more difficult to establish; for details, the reader is referred to Müller zum Hagen (1972).

Lemma 3

Suppose that the motion is Born-rigid and that $A_{[i|j]} = 0$. From the latter supposition it follows that A_i can locally be written as a gradient $\Phi_{|i}$ of some smooth function Φ. Thus, $V_{(i\|j)} = -\Phi_{|(i} V_{j)}$, and with $f = \exp(\Phi)$, it follows that $(f V)_{(i\|j)} = 0$, which means that the motion is stationary. Conversely, suppose that for some $f > 0$, $(f V)_{(i\|j)} = f V_{(i\|j)} + f_{|(i} V_{j)} = 0$. Contracting with V_j gives $f A_i - f_{|i} + f_{|j} V^j V_i = 0$. Contracting again with V^i gives $-2 f_{|i} V^i = 0$, so that $A_i = f_{|i}/f = (\log f)_{|i}$, and consequently, $A_{[i|j]} = 0$. Also using $A_i = (\log f)_{|i}$ in $0 = (f V)_{(i\|j)}$ reduces it to $V_{(i\|j)} = -A_{(i} V_{j)}$, which we saw in lemma 1 expresses Born rigidity.

Lemma 4

The proof of the Herglotz–Noether theorem is probably the hardest proof of any known fact about Minkowski space-time. For details the reader is referred to Trautmann 1965.

Lemma 5

The proof is left as a challenge to the reader.

Lemma 6

For a nonrotating Born-rigid motion $V_{i\|j} = -A_i V_j$. Consequently, $\perp V_{i\|j} \equiv k^m{}_i k^n{}_j V_{m\|n} = 0$, which says that the hypersurfaces to which V^i is orthogonal have vanishing extrinsic curvature. In flat space-time this means that the orthogonal hypersurfaces are hyperplanes.

6 Substantivalism: Newton versus Leibniz

In chapter 3, I noted that under the presupposition that a minimal form of determinism is possible for particle motions, relationist thesis (R1) (which asserts the relational character of motion) entails thesis (R2) (which asserts that space is not a substance). Had the classical relationists been aware of this implication, they would have lost no time in turning it into the obvious argument for (R2), namely, (R1), therefore by *modus ponens*, (R2). The argument is unsound, or more cautiously, the best judgment emerging from the evidence marshalled in chapters 4 and 5 is that (R1) is in fact false in both the classical and relativistic settings.

Many early participants on both sides of the absolute-relational controversy assumed that the implication went the other way round, from (R2) to (R1), the idea being that if motion is absolute rather than relational, it must take place with respect to a substantival space. This may help to explain in part why Leibniz did not use the correspondence with Clarke to respond to Newton's attack on (R1) but instead concentrated on arguing for (R2).[1] For if (R2) holds, then by the alleged implication in question, (R1) must also hold. But if (R2) does indeed entail (R1), then the relationist would lose on (R2), since (R1) fails.

Does (R2) entail (R1)? How does the failure of (R1) affect (R2)? These questions provide a major focus for the present chapter. I will begin to answer them by trying to get a better fix on what is involved in (R2) first by examining Newton's and Leibniz's pronouncements on what it means for space to be a substance and second by analyzing Leibniz's famous argument for (R2).

1 Space and Space-Time as Substances

It may come as some surprise that Newton and Leibniz agreed that in many senses space is not a substance. In the crudest sense, space would be a substance if it were a kind of stuff out of which material bodies were made. Both Newton and Leibniz would have rejected any view that smacks of a Cartesian identification of extension and matter, and each explicitly asserted that space and matter are distinct, though of course they would have put different glosses on this assertion.

Two time-honored tests for substance require that a substance be self-subsistent and that it be active. The following passage from "De gravitatione" records the result of Newton's application of these tests to space. Space "is not a substance; on one hand, because it is not absolute in itself, but as it were an emanent effect of God, or a disposition of all being; on the other hand, because it is not among the proper dispositions that denote substance, namely, actions, such as thoughts in the mind and motions in bodies" (Hall and Hall 1962, p. 132). Though we do not know what Leibniz would have made of Newton's notion that space is an emanent effect of God, he no doubt would have agreed that space fails on both counts to be a substance.

If space is not a substance, the obvious alternative is that it is an accident or property. This is the option Newton's spokesman, Samuel Clarke, endorses in his Fourth Reply to Leibniz. "Space void of body, is the property of an incorporeal substance.... Space is not a substance but a property; and if it be a property of that which is necessary, it will consequently ... exist more necessarily, (though it be not itself a substance,) than those substances themselves which are not necessary" (Alexander 1984, p. 47). Clarke's transparent suggestion here is that space is a property of God. Leibniz chose to read Clarke's further suggestion that space necessarily exists because it is the attribute of a necessary being as asserting that space is a necessary property of God, and then in his typical fashion he tries to turn the suggestion against Clarke, objecting that God will "in some measure, depend upon time and space, and stand in need of them" if they are necessary attributes of Him (p. 73).

This contretemps is in large part an irrelevant sideshow, for it is clear that at this point of the correspondence Clarke was not functioning as Newton's amanuensis but was advocating his own pet doctrine.[2] Nevertheless, the dispute does serve to raise some sticky questions about Newton's own treatment of the ontological status of space. In "De gravitatione" he claims that space "has its own manner of existence which fits neither substances nor accidents." We have seen why Newton thinks that space does not exist in the manner of substances. His reason for thinking that it does not exist in the manner of accidents uses a conceivability argument. "Moreover, since we can clearly conceive extension existing without any subject, as when we may imagine spaces outside the world or places empty of body, and we believe [extension] to exist whenever there are no bodies, and we cannot believe that it would perish with the body if God should annihilate a body, it follows that [extension] does not exist as an accident inherent in some subject. And hence it is not an accident" (Hall and Hall 1962, p. 132). Even if we accept the force of Newton's thought experiment, all that follows is that space is not an accident of bodies, and not that it is not an accident or property of God.

In his 1720 edition of the Leibniz–Clarke correspondence, Des Maiseaux's preface includes a special notice to the reader.

Since the terms *quality* or *property* have normally a sense different from that in which they must be taken here, M. Clarke has asked me to warn his readers that "when he speaks of infinite space or immensity and infinite duration or eternity, and gives them, through an inevitable imperfection of language, the name of qualities or properties of a substance which is immense or eternal, he does not claim to take the term *quality* or *property* in the same sense as they are taken by those who discuss logic or metaphysics when they apply them to matter; but that by this name he means only that space and duration are modes of existence of the Substance which is really necessary, and substantially omnipresent and eternal. This existence is neither a substance nor a quality nor a property; but is the existence of a Substance with all its attributes, all its qualities, all its properties; and place and duration are modes of this existence of such a kind that one cannot reject them without rejecting existence itself...." (Alexander 1984, p. xxix)

Koyré and Cohen (1962) have demonstrated that Newton, and not Clarke, was the author of this passage. Calling space and time "modes of existence" is not an admission that they have a quasi-property status but rather is a reference to the doctrine of "De gravitatione" that "Space is a disposition of being *qua* being": "No being exists or can exist which is not related to space in some way.... Whatever is neither everywhere [as is God] or anywhere [as are bodies] does not exist. And hence it follows that space is an effect arising from the first existence of being, because when any being is postulated, space is postulated" (Hall and Hall 1962, p. 136).

Despite all of the foregoing, it is evident that in a sense crucial to the absolute-relational controversy, Newton does take space to be a substance. Directly after claiming that the manner in which space exists fits neither substances nor accidents, he admits that "it approaches more nearly to the nature of substance": "There is no idea of nothing, nor has nothing any properties, but we have an exceptionally clear idea of extension, abstracting the dispositions and properties of a body so that there remains only the uniform and unlimited stretching out of space in length, breadth and depth" (p. 132). "Although space may be empty of body, nevertheless it is not itself a void; and *something* is there because spaces are there, though nothing more than that" (p. 138). More important than these imaginings is Newton's analysis of the role of space in the analysis of motion: "space is distinct from body" (p. 123); "place is part of space" (p. 122); "body is that which fills space" (p. 122); and motion is defined as "change of place" (p. 122). We shall shortly see that Newton's analysis of body suggests a radical reading of these remarks, but for the moment I will work with a more conservative reading. On the conservative reading, both bodies and space are substances in that bodies and space points or regions are elements of the domains of the intended models of Newton's theory of the physical world. Bodies may or may not appear in the domain of every intended model, but space points and regions do. In contrast, bodies alone exhaust the domains of the intended models of the relationist's world.

On the suggested reading, the Newtonian description of the physical state of the world at any moment has the form

$$R(p_1, p_2, \ldots, b_1, b_2, \ldots), \tag{6.1}$$

where the p_i denote points of space and the b_j denote bodies, the implicit claim being that without loss of empirical content (6.1) cannot be replaced by

$$R'(b_1, b_2, \ldots). \tag{6.2}$$

If the replaceability of (6.1) by (6.2) is to constitute a key difference between substantivalists and relationists, then both will want to agree that in (6.2) R' stands for a direct relation among bodies. Thus, both will want to reject the intermediate position, called the property view in chapter 1, which would remove space points from the ontology (the domains of the intended models) to the ideology (the predicates applied to the elements of the domain) and would translate (6.1) as

$$R''(b_1, b_2, \ldots) \;\&\; P_1(b_1) \;\&\; P_2(b_2) \;\&\; \ldots, \tag{6.3}$$

where $P_i(x)$ means that x is located at point p_i.

There are a number of ways to embellish the admittedly rather colorless characterization of substantivalism given above. A popular one is to gloss "Space is a substance" as "Space is a container for matter." It is hard to see, however, what content, if any, such a gloss adds. One may want to emphasize the container idea by saying that just as the raisins contained in a pudding can be redistributed through the pudding, so the bodies contained in space can be repositioned in space. But the latter just means that (6.1) can be true at some times but false at others, or true in some possible worlds but false in others. And this, by the way, is all that is needed to crank up the machinery of Leibniz's argument against space as a substance (see section 2 below), which gives some evidence that the relevant sense of substantivalism has been identified. That same machinery grinds away at the property view (6.3) as well as the substantivalist's (6.1).

Another embellishment is that space is ontologically prior to bodies. But again it is hard to see what this adds to the present reading of Newton. Perhaps it is meant to assert that there can be space without bodies but no bodies without space. One difficulty with this gloss comes in fixing the relevant sense of 'can'. If it is the 'can' of conceptual possibility, the substantivalist may want to accept the gloss. As we have seen, Newton certainly did, but we have also seen that by way of justification he was driven to a naked appeal to intuitions of conceptual possibility and to the slippery notion of "dispositions *qua* being." If, on the other hand, this 'can' is the 'can' of physical possibility, it may be false that there can be space without matter (i.e., in every physically possible world, bodies are included in the domain). But nothing crucial in the absolute-relational debate seems to hinge on this latter contingency.

The second embellishment comes into its own under what can be called *supersubstantivalism*, the view that space is the only first-order substance in the sense that space points or regions are the only elements of the domains of the intended models of the physical world. Thus, rather than eliminating the p_i's from (6.1), as the relationist proposes, this view eliminates b_j's to give

$$R'''(p_1, p_2, \ldots). \tag{6.4}$$

To realize supersubstantivalism, one doesn't have to revert to the view that space is stuff that forms the corpus of bodies, nor does one have to resort to some outlandish physical theory. Indeed, modern field theory is not implausibly read as saying the physical world is fully described by giving the values of various fields, whether scalar, vector, or tensor, which fields are attributes of the space-time manifold M. More will be said of the viewpoint below in chapters 8 and 9, but for present purposes it is interesting to note that Newton put forward a forerunner of such a view in "De gravitatione" as part of a tentative account of the nature of body.[3] We are asked to imagine that God has endowed various regions of space with the property of impenetrability and that impenetrability is "not always maintained in the same part of space but can be transferred hither and thither according to certain laws, yet so that the amount and shape of that impenetrable space are not changed" (Hall and Hall 1962, p. 139). Newton is careful to claim not that such beings are bodies but only that they have all the characteristics of those things ordinarily called bodies. However, he is clearly inviting the reader to entertain the hypothesis that they are bodies, in which case we can define bodies as "*determined quantities of extension which omnipresent God endows with certain conditions*" (p. 140).

Finally, I note that one traditional idea of substance holds that a substance is empowered with a means of preserving its identity through time. Newton's absolute space might seem to qualify as a substance *par excellence* on this count, since it is supposed to provide the means of identifying spatial locations through time. From the space-time viewpoint, however, this is an illusion. From this viewpoint a reference frame is just a collection of smooth timelike curves on the space-time manifold, and these curves provide the means of saying when two nonsimultaneous events have the same spatial location. To say that space is absolute in the relevant sense is just to say that some reference frame plays a special role in the laws of motion. But there is nothing in that role that confers on the special frame a magic power of sustaining its identity through time; that power is inherited from the space-time manifold, and it is inherited equally by every frame. Those who do not find the space-time point of view congenial will, of course, not be persuaded by this line. But I have tried to show in the preceding chapters that this point of view is not something foisted upon us by relativity theory but is essential to a clear understanding of classical doctrines of space, time, and motion.

2 Leibniz's Argument

Leibniz's argument against substantivalism is actually Clarke's argument turned on its head. In his Second Reply, Clarke says that he accepts Leibniz's principle of sufficient reason (hereafter, PSR), but he adds that the sufficient reason motivating God's choice may be nothing more than "mere will." "For instance: why this particular system of matter, should be created in one particular place, and that in another particular place; when, (all place being absolutely indifferent to all matter,) it would have been exactly the same thing *vice versa*, supposing the two systems (or the particles) of matter to be alike; there can be no other reason, but the mere will of God" (Alexander 1984, pp. 20–21). In his Third Letter, Leibniz in typical fashion tries to turn Clarke's argument upside down and use it as a demonstration to "confute the fancy of those who take space to be a substance, or at least an absolute being."

I say then, that if space was an absolute being, there would something happen for which it would be impossible there should be sufficient reason. Which is against my axiom. And I prove it thus. Space is something absolutely uniform; and, without the things placed in it, one point of space does not differ in any respect whatsoever from another point of space. Now from hence it follows, (supposing space to be something in itself, besides the order of bodies among themselves,) that 'tis impossible there should be a reason, why God, preserving the same situations of bodies among themselves, should have placed them in space after one certain particular manner, and not otherwise; why everything was not placed quite the contrary way, for instance, by changing East into West. But if space is nothing else, but that order or relation; and is nothing at all without bodies, but the possibility of placing them; then those two states, the one such as it now is, the other supposed to be quite the contrary way, would not at all differ from one another. Their difference therefore is only to be found in our chimerical suppositon of the reality of space in itself. But in truth the one would exactly be the same thing as the other, they being absolutely indiscernible; and consequently there is no room to enquire after a reason of the preference of the one to the other. (Alexander 1984, p. 26)

Two points of interpretation are worth mentioning at this juncture. First, Leibniz intimates that his argument was drawn from a well-stocked arsenal of confutations of substantivalism. ("I have many demonstrations, to confute the fancy of those who take space to be a substance or at least an absolute being. But I shall only use, at the present, one demonstration," p. 26) But as far as I can determine, there is nothing in the Leibniz corpus to indicate that Leibniz had other confutations up his sleeve. Nor is there any indication that this particular argument was explicitly constructed prior to the correspondence with Clarke, and the context strongly suggests that it was the product of opportunism and one-upmanship.

Second, Leibniz's operation of "changing East into West" could be read either as rotation by 180 degrees or as mirror-image reflection,[4] and the context gives no clue as to Leibniz's intentions. But whatever his intentions, there remains the question of what construction his readers put on the operation. The question is important in the case of Kant, since mirror-image reflection suggests Kant's infamous problem of incongruent counterparts. The next chapter will be devoted to a discussion of the implication of the left-right distinction for the substantivalism issue, and so I propose in the present chapter to read Leibniz's argument in terms of continuous rather than discrete transformations. To emphasize the point, I will reconstruct Leibniz's argument in terms of a shift transformation.

3 The Structure of Leibniz's Argument

I propose to understand Leibniz's objection to substantivalism as follows. Suppose that space is a substance in the sense of being an irreducible object of predication. Then a proliferation of possible worlds results, e.g.,

W, in which bodies are located in space as they now are,

W_1, in which the bodies are all shifted one mile to the east,

W_2, in which the bodies are all shifted two miles to the east,

$\vdots$

Such a richness of possibilities is an embarrassment from the point of view of the PSR, since God would be placed in the situation of Buridan's ass, with no good reason for actualizing one of the possibilities rather than another. Moreover, the proliferation of worlds also threatens to violate another cardinal principle of Leibniz's metaphysics, the principle of the identity of indiscernibles (hereafter, PIdIn). ("But in truth the one [e.g., W_{71}] would be exactly the same thing as the other [e.g., W_{38}], they being absolutely indiscernible.")

4 Leibniz's Weapons

The weapons Leibniz brandishes against the substantivalist are not quite what they seem at first glance. Depending upon the context and his interlocutor, Leibniz in his ever smooth manner glides between a causal and a theological reading of the PSR. On the former, the principle asserts that every event has a cause, or perhaps that the present state of the universe uniquely determines the future states.[5] On the latter reading, the principle refers, not to the causal nexus within an actual or possible world, but to the reasons for God's decision about which world to actualize. Of course, both readings can be covered by the gloss "Nothing happens without a sufficient reason." But such a gloss only obscures the fact that a refusal by the substantivalist to be intimidated by the bully stick of PSR does not commit him to a belief in acausality or indeterminism. The causal version of the principle will become important in chapter 9.

As for Leibniz's second weapon, the PIdIn, if it is to have the polemical force it needs, it cannot be the principle of second-order logic that goes by that name. For the substantivalist can cheerfully agree that $(P)[P(a) \leftrightarrow$

$P(b)) \rightarrow a = b]$ while at the same time maintaining that, say, W_{38} and W_{229} are distinct worlds, since W_{38} has the property that such and such bodies are in such and such region of space, while W_{229} lacks this property.[6] To respond that it begs the question to let the quantifier (P) range over absolute properties of spatiotemporal location is to seriously misrepresent the polemical situation. Newton has offered a proposal—an enormously successful proposal—and Leibniz claims to give a refutation of one key aspect of that proposal. The burden is thus on Leibniz to show why (P) cannot have such a broad scope. The burden could be discharged by establishing a version of PIdIn that limits the scope of the property quantifier to verifiable properties, thus turning the logical principle into something akin to the logical positivist's verifiability principle of meaningfulness ('A difference, to be a real difference, must be a verifiable difference'). Here as elsewhere there is an uncanny consistency to Leibniz's philosophy in that whatever reading is needed to make a coherent view can be supported by textual evidence. Thus, in his Fifth Letter to Clarke, Leibniz seems to endorse precisely the verificationism needed to make the PIdIn prong of his argument work.—"Motion does not indeed depend upon being observed; but it does depend upon being possible to be observed. There is no motion, when there is no change to be observed. And when there is no change that can be observed, there is no change at all" (Alexander 1984, p. 74).

Although the textual evidence indicates that Leibniz did intend a verificationist interpretation, it is nevertheless interesting to inquire whether his argument can be given a nonverificationist twist. Teller thinks that it can.[7] "Except for the alleged differences in space-time placement, there is *no* difference—observational, theoretical, or whatever—between the alternatives we are considering.... But if two descriptions agree in such a thoroughgoing way, surely sound methodology and good sense require us to count them as verbally different descriptions of the same situation" (1987, p. 433). The substantivalist will see no force at all to this version of the argument. He has posited space or space-time points not on whim but because be believes that such a posit is necessary for a well-founded theory of motion. Thus, naked appeals to "sound methodology" and "good sense" are not going to sway him, since, he claims, sound methodology and good sense were what led to the substantivalist account of space. If there are specific objections to the methodology and good sense of Newton and fellow substantivalists, then they should be made at the appropriate places.

I conclude that if Leibniz wants to carry through his avowed aim of refuting those who take space to be a substance, as he claims in his Third Letter to Clarke, he must, in the absence of any constructive alternative to Newton, stick to the verificationist version of his construction.

5 The Reach and Implications of Leibniz's Argument

Although the announced purpose of Leibniz's argument is to establish the relationist thesis (R2), the argument works just as well (or ill) to establish a form of thesis (R3). For trading an ontology of space points for an ideology of irreducible nonrelational properties of spatial location leads to a parallel proliferation of possible worlds that also grates against the theological version of PSR and the verifiability version of PIdIn.

Had there not been so much bad blood between Leibniz and the Newtonians, Leibniz would no doubt have noted that his argument lends itself to the ecumenicalism and reconciliationism that motivates much of his work.[8] For the argument shows the substantivalist the error of his ways, while at the same time suggesting how part of substantivalism can be preserved. If it works, Leibniz's argument shows that the substantivalist provides a phony picture of physical reality, but the phoniness is not that of a doctored photograph that shows the cat on a mat when it is really on the sofa. Rather, the substantivalist picture provides an accurate rendering or representation of reality, but the representation relation is one-many, with many (indeed, uncountably many) substantivalist pictures corresponding to the same relationist reality. Mistakenly thinking that the correspondence between his picture and reality is one-to-one rather than many-one leads the substantivalist into problems with determinism (see section 3.6), with PSR, and with PIdIn. This representational ploy will be exploited in chapter 7 in the discussion of incongruent counterparts and will be examined in more technical detail in chapters 8 and 9.

The ploy does appear to have an unpalatable consequence. Substantivalism sanctions an active picture of space-time and dynamic symmetries: a symmetry transformation of the space-time acts on a system of particles to produce, for example, a rotation or translation of the system in space or a velocity boost of the system. The transformation is a true symmetry of nature just in case the history of the system resulting from applying the transformation is physically possible whenever the initial history is. The suggested reading of relationism rejects the active picture of symmetry transformations and enforces instead a passive reading: a dynamic-symmetry transformation connects not different physical systems that are repositioned or reoriented in space but rather different descriptions of the same system. But if all these descriptions are equally accurate, it would seem that the symmetry transformation could not fail to be a true symmetry of nature, contradicting the usual understanding that symmetry principles are contingent, that is, are true (or false) without being necessarily true (or false). On further reflection, however, the relationist is no more committed to the necessity of symmetry principles than is the substantivalist. Either a physical theory satisfies the metasymmetry principles (SP1) and (SP2) of chapter 3 or not. If not, the theory is ill-formulated. If so, the laws of the theory must exhibit the symmetries of the space-time in which they are based. This is true for both the substantivalist and the relationist; where they part company is over what symmetries space-time should exhibit (essentially, issue [R1]) and over the ontological status of space-time (issue [R2]).

6 Limitations (?) on Leibniz's Argument

In an otherwise lucid discussion of relationism, Friedman (1983, chapter 6) opines that if matter forms a plenum so that there are no unoccupied space points, then the traditional dispute over (R2) would be moot. Leibniz would have been surprised at this opinion, for he believed in a plenum and yet he certainly did not think that this belief rendered his relational conception of space functionally equivalent to that of the absolutist. A good test to apply to this case is to ask how Leibniz's argument fares in the case of a plenum. It seems that to the extent that the argument works at all, it works equally well against the background of a plenum as against the background of a discrete system of particles moving in a void. The root of Friedman's misdirection is the notion that in a plenum the relationist's ontology would be just as rich as the substantivalist's. This matter will be taken up in chapter 8.

A number of commentators trace another potential limitation to Leibniz's assumption that space is "something absolutely uniform; and, without the things placed in it, one point of space does not absolutely differ in any respect whatsoever from another point of space," or in modern jargon, space is assumed to be homogeneous and isotropic. The mathematical apparatus for stating coherent alternatives to this assumption was unavailable in the formative period of the debate, but nothing prevents us from entertaining the possibility that space (or space-time) is, say, variably curved.[9] There are still two subsequent suppositions that can be made. First, the nonuniform structure of space (or space-time) might be immutable in the sense of being the same in every physically possible world. In that case both the PSR and PIdIn prongs of Leibniz's argument can be blunted: God might have good reason to place bodies in a region of low rather than high curvature (say, because the former is a more stable configuration), and shifting bodies from a region of one curvature to a region of a different curvature could in principle be experimentally detected (say, because the motion of gravitating bodies is lawfully connected with the curvature of space). Second, the nonuniform-curvature structure might not be immutable but might, as in GTR, covary with the distribution of matter. That in itself might or might not be heralded as a victory for relationism (see chapter 5), but the implications for Leibniz's argument are not immediately evident. One cannot now coherently speak of shifting bodies from a region of high curvature to a region of low curvature if mass creates curvature. However, if the underlying laws are such that shifting the bodies drags along with it the curvature structure, then the PSR and PIdIn arguments would seem to be reactivated.

This last observation also necessitates a reappraisal of what happens under the first supposition. For as we have seen in chapter 3, any space-time theory that satisfies some mild restrictions admits of a simultaneous dragging along of the space-time and matter structures on the space-time manifold, whether or not the space-time structures are immutable. The moral, to be explored in more detail below in section 8, is that whether or not nonuniformity of space-time structure blocks Leibniz's argument depends upon what form of substantivalism the argument is supposed to refute. But before exploring this point, I need to examine various responses to Leibniz's argument.

7 Responses to Leibniz's Argument

The refusenik strategy

The refusenik simply refuses to engage in the kind of argumentation Leibniz uses, the basis for his refusal being the protest that he does not understand the modal discourse it presupposes and especially the notion of "possible worlds" (see Field 1985). Of course, there is nothing Leibniz can do to budge an opponent who simply refuses to play the game, but he can make the refusenik look like an unworthy opponent. To this end he can point out that his argument doesn't need to use possible worlds conceived as foreign planets or concrete objects. A substantivalist theory of physics will admit an array of models; substantivalist laws of motion will admit an array of solutions; and the argument can be run on these arrays. To refuse to engage in this form of argumentation is to refuse to do science as we know it.

The blunting strategy

Blunt the PIdIn prong of the argument Recall from section 4 that what is needed to make this prong work is not the logical principle that goes by the name of PIdIn but a form of verificationism. In this postpositivist era it is not unrespectable to reject verificationism, and so there need be no great embarrassment in maintaining that there are ontologically distinct but epistemologically indistinguishable worlds or situations (see Horwich 1978).

Blunt the PSR prong Two forms of this substrategy can be contemplated. We saw that PSR divides into causal and theological versions, the latter of which is needed for Leibniz's argument. The substrategy could be carried out by rejecting the latter version *in toto* or at least rejecting the part that tells a creation story of actuality. What, after all, is this mysterious property of actuality that God is supposed to confer upon a world when he chooses to actualize it? Avoid the question by adopting instead an indexical account of actuality (see Lewis 1986).

Alternatively, the substrategy can be implemented by melding Leibniz's creation story of actuality with Clarke's account of divine decision making. Clarke held that if God could not act from mere will without a predetermining cause, "this would take away all power of choosing, and ... introduce fatality" (Alexander 1984, p. 25). Leibniz responded that Clarke grants PSR "only in words, and in reality denies it." Clarke's view involves "falling back into loose indifference which ... [is] absolutely chimerical even in creatures, and contrary to the wisdom of God" (p. 27). Clarke rejoined that "to affirm in such case, that God cannot act at all, or that 'tis no perfection in him to be able to act, because he can have no external reason to move him to act one way rather than the other, seems to be denying God to have in himself any original principle or power of beginning to act, but that he must needs (as it were mechanically) be always determined by things extrinsic" (pp. 32–33). Leibniz responded that moral necessity "does not derogate from liberty" (p. 56), and Clarke rejoined that a truly free agent, even when faced with "perfectly alike reasonable ways of acting, has still within itself, by virtue of its self-motive principle, a power of acting" (p. 98). This aspect of the debate ends in an inconclusive stalemate.

Horwich's strategy

Consider electrons as objects of predication, and let C be a list of the characteristics an electron may possess (mass, spin, etc., and, if you like, spatial position). Start with a world W where electron A has characteristics C_A and electron B has characteristics C_B. Now produce another world W' just like W except that A now has characteristics C_B while B has C_A. We would hardly let PSR and PIdIn arguments as applied to W and W' convince us that electrons do not exist (see Horwich 1978). So why should we let them convince us that space points do not exist?

The point can be more generally put this way. No theory written in any standard logical language, whether the underlying logic is first order or *n*th order, can fix the models more closely than up to isomorphism. Thus, the existence of isomorphic models does not in itself cast doubt on the adequacy of the theory or the existence of the entities postulated by the theory. Of course, there may be something suspect about some particular class of isomorphic models, but the case has to be made on the particulars. In chapter 9, I shall make the case in terms of the possibility of determinism in the context of general relativity and similar theories.

Field's strategy

Horwich's construction can be seen as an aid to the blunting strategy; in particular, it could be used to buttress the rejection of the idea that epistemologically indiscernible worlds must be identical. Alternatively, it can be seen as a motivation for resisting the relationist's charge that substantivalism gives rise to an embarrassing proliferation of worlds. Using Leibniz' s PIdIn across worlds, we could conclude that the electrons exhibiting C_A and C_B in W' are respectively A and B, and thus that W and W' aren't distinct worlds after all. Field (1985) suggests that the substantivalist apply a similar strategy to space points. This is not a strategy that Newton and other traditional substantivalists would have wanted to use, since they never thought that the identity and individuation of space points turn on what material objects occupy which points. Indeed, the supersubstantivalism suggested explicitly in Newton's "De gravitatione" presupposes that space points are individuated independently of the matter fields on space. Moreover, the notion that identification across possible worlds respects isomorphism is incoherent if the worlds exhibit such symmetries that multiple isomorphisms exist. Field suggests that transworld identification should respect *some* isomorphism, but he gives no indication of what considerations would pick out the right one when many exist. We will have to wrestle with these mysteries of predication and individuation again in chapter 9.

Two provisional conclusions can be drawn from our discussion to this point. First, only dyed-in-the-wool absolutists would deny that Leibniz's argument has considerable intuitive appeal. Second, only dyed-in-the-wool relationists would claim that the argument has enough polemical force to do what Leibniz asserted it would do, namely, refute substantivalism, for there are reasonable positions open to Leibniz's opponents that will blunt both prongs of the argument. In chapter 9, I will try to show that considerations of determinism in theories that abandon the immutability of space-time structure add enough polemical force to Leibniz's argument to make it much more interesting, if not fully persuasive.

8 The Absolutist Counterattack

In a nutshell, the absolutist attack consists of first noting that (R1) fails and then claiming that the failure carries with it a failure of (R2). The original Newtonian form held that (R1) is false, because the analysis of motion requires the concept of absolute change of position, which concept has to be understood as change of position with respect to a substantival space. The modified form concedes that the scientific treatment of motion doesn't require absolute change of position or absolute velocity but asserts that it does require some absolute quantities of motion, such as absolute acceleration or rotation. To make these quantities meaningful requires the use of inertial structure or the like, and these structures must be properties of, or inhere in, something distinct from bodies. The only plausible candidate for the role of supporting the nonrelational structures is the space-time manifold M, used extensively in the discussion of space-time theories in preceding chapters.

Note that this manifold substantivalism—M as a basic object of predication—lays itself open to Leibniz's argument whether or not space or space-time is "absolutely uniform." Consider a substantivalist model $\mathcal{M} = \langle M, A_1, A_2, \ldots, P_1, P_2, \ldots \rangle$, where the A_i and P_j are object fields on M characterizing respectively the space-time structure and the physical contents of space-time. Think of a diffeomorphism $d: M \rightarrow M$ as a kind of Leibniz shift. As a generalization of Leibniz's argument, let d shift the space-time structure as well as the physical contents i.e., the shift operation changes $\mathcal{M}$ to $\mathcal{M}^d = \langle M, d * A_1, d * A_2, \ldots, d * P_1, d * P_2, \ldots \rangle$. If we assume general covariance of the laws of the theory (see chapter 3), $\mathcal{M}^d$ will be a model of the theory if $\mathcal{M}$ is. The PSR and PIdIn pincers can now be applied to $\mathcal{M}$ and $\mathcal{M}^d$, irrespective of whether or not d is a symmetry of the A_i.

But this generalized version of Leibniz's argument is no more polemically powerful than the original version; the same responses discussed in section 7 apply here as well. Since the absolutist can point to a good reason for accepting manifold substantivalism and since Leibniz's argument is not powerful enough to force a retreat, on balance the absolutist seems to win on (R2) as well as (R1). The following section discusses a clever but desperate attempt to avoid this conclusion.

9 Sklar's Maneuver

On behalf of the antisubstantivalist who wishes to resist the move from the failure of (R1) to the failure of (R2), Sklar (1976, pp. 229–232) contemplates a maneuver so clever (and perhaps so outlandish) that it never occurred to any of the parties to the debate.[10] In brief, the idea is to give up on (R1) and concede that an adequate theory of motion must employ, say, absolute acceleration but at the same time to defend (R2) by treating absolute acceleration as a primitive property of particles. This primitive property is monadic. To say that a particle is absolutely accelerated in this new sense is not to make any relational claim at all, and in particular, it is not to claim that the particle is accelerating either relative to Mach's fixed stars or relative to immaterial Newtonian reference frames; rather, it is to make a claim analogous to saying that the particle is red or is massive.

One qualm about Sklar's maneuver can be brushed aside. One might worry that the maneuver seems to violate the spirit, if not the letter, of relationist thesis (R3), which prohibits the use of monadic spatiotemporal properties (see chapter 1). But this violation need not cause the relationist any consternation, for the contemplated use of the monadic property of absolute acceleration, unlike the use of such monadic properties of spatial location as 'is located at space point *p*', does not give rise to a proliferation of possible worlds that cranks up Leibniz's PSR and PIdIn objections.[11]

Before any further evaluation of the maneuver can be attempted, we need to get a better fix on exactly what is being claimed. Sklar is not offering a constructive, third alternative to standard relational and absolute theories. Nor, presumably, is he advocating some cheap instrumentalist rip-off of Newtonian theory. The only remaining alternative on the horizon is to dovetail the maneuver with the reconciliationist interpretation of Leibniz's argument (section 5 above) by exploiting the idea that absolutist models provide representations of physical reality and by taking neo-Newtonian acceleration as a representation of Sklar's primitive absolute acceleration. The idea is intriguing, but the details are missing in Sklar's account, and in what follows I shall have to speculate about how they might be filled in so as to overcome two obvious difficulties.

The first is that the representation relation cannot obtain as a matter of magic. If all Sklar's antisubstantivalist can say about his primitive acceleration is that it is the quantity represented as neo-Newtonian absolute acceleration, then the maneuver amounts to no more than hand waving. Notice that there is no corresponding problem here for the honest Leibnizian relationist. A Leibniz state is specified by listing the particles and their relative distances, relative velocities, etc. An absolutist representation is obtained by embedding the particles in Leibnizian space-time in such a way that the relative particle quantities are respected. There are, of course, many such embeddings—that is the point of Leibniz's argument. There is a problem, however, when the embedding is into neo-Newtonian space-time, to which we seem to be driven in order to get an adequate theory of motion. To solve the problem, Sklar's antisubstantivalist must characterize his primitive acceleration in such a manner that does not foster substantivalism but still fixes in some natural and perspicuous way the intended representation. Toward this end, one could stipulate that in the gravitational case the Sklar acceleration of particle *i* is characterized by a non-negative number and direction that are equal respectively to the magnitude and direction of $\sum_{j \neq i} m_j \hat{\mathbf{r}}_{ij}/r_{ij}^2$. This serves to ground the representation of the Sklar acceleration in neo-Newtonian space-time as a spacelike four-vector. But it remains magic that the representative is neo-Newtonian acceleration $(d^2x^i/dt^2)/dt^2 + \Gamma^i_{jk}(dx_j/dt)(dx_k/dt)$ or for that matter that it is a mechanical quantity at all. (When a particle ontology is traded for a field ontology, the matter becomes more desperate, since the very characterization of fields seems to require reference to the space-time manifold. This suggests that the manifold substantivalist has a separate motivation that proceeds independently of a denial of [R1]. This suggestion will be examined in chapters 8 and 9.)

The second shortcoming of Sklar's proposal is that it says nothing about how to use the envisioned nonsubstantivalist vocabulary to formulate principles of motion that can be used to explain and predict particle trajectories. Of course, the Newtonian apparatus can be used to make the predictions and afterwards discarded as a convenient fiction, but this ploy is hardly distinguishable from instrumentalism, which, taken to its logical conclusion, trivializes the absolute-relational debate, as it does so many issues in the philosophy of science.

This second problem is not really separable from the first. Sklar's vocabulary must be rich enough to code recognizable analogues of the Newtonian laws of motion, and the analogues must be close enough that one can see that in some natural way the representations of a given Sklar model are all appropriately equivalent Newtonian models. That Sklar's hypothetical vocabulary can be rich enough to achieve these goals and not at the same time introduce close analogues of the very substantival ontology he wishes to avoid is at this juncture only a pious hope. In chapters 8 and 9, I discuss ways in which this hope can be partially realized, but I emphasize that only a partial realization is on the horizon.

Whatever its merits for furthering the absolute-relational controversy, Sklar's maneuver might be thought to help rationalize Leibniz's puzzling doctrine of "force."[12] Before taking up this matter in section 11, I shall consider another relationist maneuver that seeks to take advantage of the relativistic nature of space-time.

10 Another Relationist Maneuver

Although relativity theory may make it more difficult for the relationist to maintain (R1) (see chapter 5), it may make it easier for him to resist the pressure to move from a failure of (R1) to a failure of (R2). The reason for the resistance has two facets. First, in contrast to the classical space-times studied in chapter 2, a relativistic space-time employs a space-time metric,

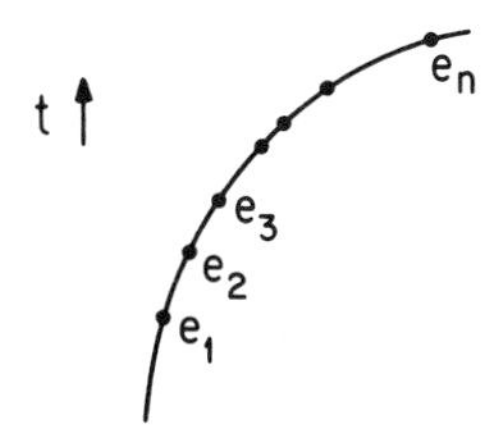

Figure 6.1
A relational account of absolute acceleration in Minkowski space-time

and all the structures of space and time derive from this metric. Second, in the classical setting, where an absolute notion of simultaneity is available, it is natural for the relationist to concentrate on such instantaneous relations as the distance between two particles at a given time, their instantaneous relative speed, etc. In contrast, the relativistic space-time setting suggests that the relationist may avail himself of a much richer set of relations. To see how the relationist might take advantage of these facts, consider how he might treat motion in Minkowski space-time.[13]

A set of particle events is specified along with a relation of genidentity and an assignment to any pair of events e_1 and e_2 of a real number $\Delta s(e_1, e_2)(= \Delta s(e_2, e_1))$. In addition, if $\Delta s(e_1, e_2) > 0$, the ordered pair (e_1, e_2) is assigned a $+$ or a $-$. This relationist description is to be represented as follows. The particle events become points on timelike world lines in Minkowski space-time, with two events lying on the same world line just in case they are genidentical. The expression $\Delta s(e_1, e_2)$ represents the Minkowski interval between e_1 and e_2, and the $+$ $(-)$ means that the direction from e_1 to e_2 is future (past) directed. One then goes on to prove a representation theorem to the effect that the relationist description determines the particle motions, since, except in special circumstances, the representation is determined up to a Poincaré transformation. (This follows from the fact that if enough independent points in Minkowski space-time are given, the Minkowski distances to another point suffice uniquely to fix the other point.[14])

Note that this defense of (R2) concedes that (R1) is false, since, for example, the relationist description of a single isolated particle will determine whether it is accelerating. Thus, for example, the relationist knows that the particle pictured in figure 6.1 is accelerating, since the Minkowski distance from e_1 to e_n is less than the sum of the distances from e_1 to e_2, from e_2 to $e_3, \ldots$, and from e_{n-1} to e_n. The modern relationist can seek to

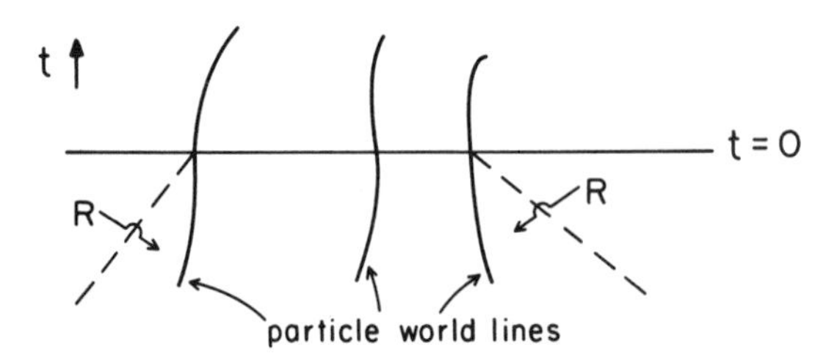

Figure 6.2
Relationism and determinism for particle motions

justify the seeming apostasy in abandoning (R1) by noting that of the two main motivations for (R1)—namely, verificationism (see chapter 3) and the seeming need to maintain (R1) as a necessary condition for (R2)—the first is no longer attractive and the second is mistaken.

GTR is much less friendly to the sort of relationist ploy under discussion. The Minkowski interval between e_1 and e_2 is the space-time length of the unique geodesic connecting these events. In general-relativistic space-times an arbitrary pair of events may not be connected by a geodesic, and even if it is, the geodesic may not be unique. I waive this difficulty to state a deeper one.

Suppose that we are only concerned with explaining particle motions. The relationist must hold that talk about space-time itself, as opposed to spatiotemporal relations among particle events, can only be construed as a convenient way of representing relational facts about particles. If this is so, it should at least be a live possibility that relational facts about the past motions of the particles determine their future motions. But in general-relativistic space-times this will not be so for a finite system of particles. For facts about the past motions of the particles in figure 6.2 will not suffice to fix the state of the space-time outside of the region R, as the future motions of the particles will in general be influenced by the nature of the space-time metric to the past of $t = 0$ and the future of R. To succeed, the relationist must resort to a version of Sklar's maneuver as applied to the metric field in particular and to physical fields in general. The prospects for success will be discussed in chapters 8 and 9.

11 Leibniz on Force

While still denying the "absolute reality of space" in his Fifth Letter to Clarke, Leibniz makes what appears to be a concession: "I grant that there is a difference between an absolute true motion of a body, and a mere relative change of its situation with respect to another body. For when the immediate cause of the change is in the body, that body is truly in motion; and then the situation of other bodies, with respect to it, will be changed consequently, though the cause of that change be not in them" (Alexander 1984, p. 74). Clarke took this to be a damning admission: "Whether this learned author's being forced here to acknowledge the difference between absolute real motion and relative motion, does not necessarily infer that space is really a quite different thing from the situation or order of bodies; I leave to the judgment of those who shall be pleased to compare what this learned writer here alleges, with what Sir Isaac Newton has said in the *Principia*, Lib. I, Defin. 8"[15] (p. 105). Modern commentators who take up Clarke's invitation also tend to see Leibniz's admission as damning. Thus, in his introduction to the correspondence H. G. Alexander allows that "There is ... no doubt that this admission of the distinction between absolute and relative motion is inconsistent with his general theory of space" (Alexander 1984, p. xxvii).

Let us try to be a little more precise about the snare Leibniz seems to have constructed for himself. When Leibniz speaks of the "immediate cause" of the change of situation of a body, he is referring to "force," his measure for which is *vis viva* or mass $\times$ (velocity)^2. Thus, to say that a body is truly in motion is on Leibniz's account to say that the velocity of the body is nonzero. But for such talk to be meaningful it would seem that space-time must have at least as much structure as full blown Newtonian space-time (see section 2.5 above), with absolute space.

On Leibniz's behalf one might insist on distinguishing between absolute space as a special reference frame and absolute space as a substantival container for bodies. Leibniz's main target in the polemic with Clarke is the latter. This is certainly true of his Third Letter, and paragraph 53 of his Fifth Letter, where he distinguishes between true and relative motion, begins by saying that he finds nothing in Newton's Scholium "that proves, or can prove, the reality of space in itself," which presumably refers to space as a substantival entity. But no amount of logic chopping will restore the appearance of consistency. For even if we grant that Leibniz's main aim was to secure (R2), there remains the fact that even by Leibniz's lights (R2) entails (R1), and so his admission, which seems to contradict (R1), is inconsistent with (R2).

A clue to the origin of Leibniz's bind is to be found in the 1694 correspondence with Huygens, a correspondence in which both parties were at pains to declare that they had found sufficient grounds to reject Newton's attempt to use rotation as a justification for absolute space (see chapter 4). In the midst of one such declaration Leibniz adds:

But you will not deny, I think, that each body does truly have a certain degree of motion, or if you wish, of force, in spite of the equivalence of hypotheses about their motions. It is true that from this I draw the conclusion that there is something more in nature than what geometry can determine about it. This is not the least important of the many arguments which I use to prove that besides extension and its variations, which are purely geometrical things, we must recognize something higher, namely, force. (Loemker 1970, p. 418)

The last sentence indicates that Leibniz felt compelled to swallow whatever consequences flow from his doctrine of force, for it is that doctrine that forms the keystone of his rejection of the Cartesian analysis of substance in terms of purely "geometrical" quantities.

In "A Brief Demonstration of a Notable Error of Descartes and Others concerning Natural Law" (1686), Leibniz claimed that force is to be estimated by the effects it can produce and that it follows on this estimate that the amount of force is equal not to Descartes's quantity of motion, mv, but to mv^2. These claims are repeated in section 17 of the "Discourse on Metaphysics" (1686). Section 18 adds the claim that this *vis viva* leads to a better understanding of the principles of motion:

For considering only what it means narrowly and formally, that is, a change of place, motion is not something entirely real; when a number of bodies change their position with respect to each other, it is impossible, merely from a consideration of these changes, to determine to which of the bodies motion ought to be ascribed and which should be regarded at rest.... But the force or the immediate cause of these changes is something more real, and there is a sufficient basis for ascribing it to one body rather than to another. This, therefore, is also the way to learn to which body the motion preferably belongs. (Loemker 1970, p. 315).

Taken at face value, this passage seems to be claiming that there is in principle a way to determine to which bodies the force belongs and, therefore, to determine which bodies are in true motion. As such, the passage would amount to a straightforward denial of the thesis (R1), which asserts the relational character of motion.

H. G. Alexander tries to put a better face on what seems to be an inconsistent performance.

Leibniz does not say [in the fifth letter to Clarke] whether it is ever possible in practice to determine in which of several bodies the cause of their change of relative position lies and so to discover which is truly in motion. He may therefore have held that the distinction between absolute and relative motion is metaphysical, not physical: that is, the absolute motion of a body can never be experimentally determined; and so the concept of absolute motion is of no use in physics. Such an interpretation is supported by his statement in the *Discourse on Metaphysics*, that moving force is a metaphysical concept. (Alexander 1984, pp. xxvi–xxvii)

But it is clear that when Leibniz calls force a metaphysical concept, he does not intend 'metaphysical' as a contrast to 'physical,' for at the beginning of section 18 of the "Discourse" he says that force as distinguished from quantity of motion is important in "physics and mechanics in finding the true laws of nature and the rules of motion."[16] Rather, Leibniz uses 'metaphysical' as applied to force to contrast with the Cartesian geometrical quantities of size, figure, and motion and also to indicate the teleological character of force as a quantity to be estimated by its effects. Moreover, the clear impression given by section 18 of the "Discourse" is that it is possible to discover which body is truly in motion.

Alexander is correct, however, that eight years later in the correspondence with Huygens, Leibniz gives exactly the opposite impression: "Even if there were a thousand bodies, I still hold that the phenomena could not provide us (or angels) with an infallible basis for determining the subject or the degree of motion and that each body could be conceived separately as being at rest" (Loemker 1970, p. 418). Perhaps realizing the awkward implications of his doctrine of force, Leibniz now seems to want to retreat. But has he left himself any consistent ground to which to retreat?

The essay "Phoranomous, or, On Power and the Laws of Nature," tentatively dated 1688, that is, after the "Discourse" and before the correspondence with Huygens, can be seen as an attempt to provide that ground. The essay begins with the assertion that because of (what we would call) Galilean invariance of the laws of impact, "not even an angel" could determine which bodies in a system of bodies moved by collisions are at rest. It continues:

To summarize my point, since space without matter is something imaginary, motion, in all mathematical rigor, is nothing but a change in the situations of bodies with respect to one another.... But since, nonetheless, people do assign motion and rest to bodies ..., we must look into the sense in which they do this, so that we do not judge them to have spoken falsely. And we must reply that one ought to choose the more intelligible hypothesis and that there is no truth in a hypothesis but its intelligibility.[17] (Couturat 1903, pp. 590–591)

But what is immediately disconcerting about Leibniz's discussion here is the absence of any reference to force.[18] And brushing aside that qualm and applying the formula that truth equals intelligibility to force leads to a dilemma. On one reading of the formula, there is really no truth to the matter as to what bodies possess force, and thus we are free to choose the hypothesis about motion that we find most intelligible. This reading squares with some of the sentiments in the "Phoranomous": "And since one hypothesis might be more intelligible than another from a different point of view with respect not so much to men and their opinions, but with respect to the very matters at hand, and thus one might be more appropriate than another for a given purpose, so also from a different point of view the one might be true and the other false" (p. 591).

But such a reading doesn't square with the sentiments of the "Discourse" and the correspondence with Huygens and Clarke, where Leibniz's insistence on the distinction between true and relative motion contains no hint that the dividing line can change with one's point of view or purpose. On an alternative reading of the formula, there is a truth to the matter and that truth consists precisely in intelligibility, or as Leibniz puts it elsewhere, in simplicity. Consider the hypothesis that, say, bodies *a*, *c*, *e* possess force and are truly in motion, while *b*, *d*, and *f* have no force and are at rest. If this is true because it is the simplest and most intelligible hypothesis, then it becomes a mystery why neither we nor angels can determine its truth, since, presumably, we and our more angelic forms do have access to considerations of simplicity and intelligibility.

Finally, Leibniz might avail himself of Sklar's maneuver and declare that force is a property of particles that is primitive and monadic and that manifests itself in absolutist representations as *vis viva*. But presumably such a representation is correct if and only if it assigns a nonzero *vis viva* just in case the force is nonvanishing. And with this structure, all of the above problems arise again at the level of representations.

12 Possibilia and Relationism

Possibilia have always been part of the discussion of relationism, as evidenced by Leibniz's pronouncement that "space denotes, in terms of possibility, an order of things which exist at the same time" (Alexander 1984, p. 26). My own reading of this dictum assigns a relatively benign role to possibilities; namely, Leibniz meant only to indicate that bodies may stand in Euclidean relations to one another in many different configurations and that each such configuration can be represented in many ways by embeddings into a container space. But other commentators, even those not predisposed toward relationism, have averred that the relationist can make himself proof against defeat if he is willing to avail himself of *possibilia* in some stronger sense. This sentiment is one of the most pernicious in the entire field, for it serves to obscure the substantive aspects of the absolute-relational debate.

The relationist can certainly make himself proof against attack from the absolutist by taking talk about space or space-time points seriously but giving such talk a relationist gloss. He does so by taking it to be talk about permanent possibilities of location for bodies or events. This defense measure succeeds only by eroding the difference between relationism and substantivalism; indeed, the notion that space points are permanent possibilities of location for bodies is one plausible reading of the substantivalism of Newton's "De gravitatione."

Alternatively, the relationist can try to keep his doctrine distinct by keeping space or space-time points at arm's length while at the same time providing a relational transcription of talk about space and space-time. Thus, if it becomes necessary to talk about the state of unoccupied regions of space and space-time, the relationist can substitute talk about how hypothetical particles would behave in relation to each other and in relation to actual particles were they to be introduced into the appropriate regions. Presumably the behavior in question refers to nonaccidental, lawlike features of particle motions, and if so, the subjunctive and counterfactual conditionals must ultimately find their licence in laws of physics. But the absolutist claims that the laws of physics cannot be stated without the use of an apparatus that carries with it a commitment to substantivalism. In this the absolutist may be wrong, but he is right in holding that the relationist cannot take refuge in subjunctive and counterfactual talk and must meet the challenge of formulating a relationally pure physics. The history of relationism is notable for its lack of success in meeting this challenge. Finally, the empiricist tradition, from which most relationists come, holds that laws of nature, dispositions, and potentialities all supervene on the actual, occurrent facts (see chapter 5 of my 1986). If correct, this means that there is no real difference between a conservative relationist, who holds that all there is to the world are such facts as that body b_1 is five meters from body b_2 at time t, and a liberal relationist, who holds that there are also subjunctive facts about what would happen if, say, b_3 were introduced between b_1 and b_2.

13 Conclusion

The provisional assessment of (R2) is necessarily more tentative and vague than our assessment of (R1), first because it is less clear what is at stake in (R2) and second because the arguments pro and con on (R2), while ingenious and intriguing, are far from compelling. On balance, however, the relationist seems to be on the defensive in trying to maintain that space or space-time is not a substance. Leibniz's argument for (R2) undeniably has an intuitive tug, but its force is far from irresistible, since there are plausible responses to each of its two main prongs. Furthermore, (R1) has been discredited, and the discredit seems to transfer, *pace* Sklar, to (R2). The next chapter examines an attempt by Kant to further tip the balance against the relationist.

7 Kant, Incongruent Counterparts, and Absolute Space

In his 1768 essay "Concerning the Ultimate Foundation of the Differentiation of Regions in Space," Kant used incongruent counterparts in an attempt to refute a Leibnizian-relationist account of space. It is hard to imagine that scholars could be more divided on how to understand Kant's argument and on how to assess its effectiveness (compare Alexander 1984/1985; Broad 1978; Buroker 1981; Earman 1971; Gardner 1969; Lucas 1984; Nerlich 1973, 1976; Remnant 1963; Sklar 1974; Van Cleve 1987; Walker 1978; and Wolff 1969). Two years later in 1770 incongruent counterparts resurface in Kant's *Inaugural Dissertation*, this time as part of a proof that our knowledge of space is intuitive. They appear yet again in the *Prolegomena* (1783) and in the *Metaphysical Foundations of Natural Science* (1786) as part of the argument for transcendental idealism. Not surprisingly, scholars are also at odds on how to explain the shifts in the roles Kant wanted incongruent counterparts to play and on how to assess the importance of these matters for the development of his critical philosophy (compare Alexander 1984/1985; Allison 1983; Bennett 1970; Broad 1978; Buchdahl 1969; Buroker 1981; Walker 1978; Winterbourne 1982; and Wolff 1969).

The present chapter is concerned primarily with Kant's 1768 argument. It aims at both a better understanding of Kant's argument and a sharper formulation of the ways incongruent counterparts are and are not relevant to the controversy over absolute versus relational conceptions of space. A few remarks are offered on Kant's post-1768 use of incongruent counterparts.

1 Kant's Argument against Relationism

Although Kant is his usual cryptic self in "Concerning the Ultimate Foundation of the Differentiation of Regions in Space," an argument against Leibniz's relational conception of space is readily extractable.

K1 "Let it be imagined that the first created thing were a human hand, then it must necessarily be either a right hand or a left hand." (1768, p. 42)

It follows, supposedly, that the relational theory is not adequate, since

K2 on the relational theory the first created thing would be neither a right hand nor a left hand since the relation and situation of the parts of the hand with respect to one another are exactly similar in the cases of right and left hands that are exact mirror images of one another.

Before I turn to an evaluation of Kant's argument, it is worth remarking on how novel it was. As we have seen above, the standard absolutist attack on relationist thesis (R2) was to claim that motion is absolute rather than relational and that absolute motion must be understood as motion with respect to a space that is ontologically prior to bodies. Kant thought that he could arrive at the same result—that, as he puts it, "*absolute space has its own reality independently of the existence of all matter*" (1768, p. 37)—by showing that the differences between right and left "are connected purely with *absolute and original space*" (p. 43). Kant was, of course, aware of the more standard objections to relationism; indeed, he specifically mentioned Leonhard Euler's attempt to show that the relational theory cannot properly ground Newton's first law of motion,[1] but he brushed it aside as having a merely *a posteriori* character. Kant's aim was to "place in the hands, not of engineers, as was the intention of Herr Euler, but in the hands of geometers themselves a convincing proof" of the reality of absolute space. (pp. 37–38).

Let us turn now to Kant's argument itself. The relationist may wish to attack either (K1) or (K2). Consider first the attack on (K1). The relationist may urge that whether or nor a hand is right or left depends upon the relation of the hand to an appropriate reference body. Thus, contrary to Kant, for a hand standing alone there aren't two different actions of creative cause for God to choose between.[2] If a reference body is introduced, then as the relationist will readily agree, different acts of creative cause are required for the hand in question to have different relations to the reference body.

It is instructive to compare the situation here to that for continuous symmetry transformations. According to the absolutist, the operation of shifting all the bodies one mile to the east in the container space produces a different state of affairs, and different acts of creative cause are required for God, according to which of these situations He chooses to actualize. As we saw in chapter 6, Leibniz charged that this result violates the principle of sufficient reason, since God would have no good reason to actualize one rather than another of these absolutist states, and he proposed to rescue God from the situation of Buridan's ass by maintaining that what the absolutist is providing is different descriptions of the same intrinsic (relational) state of affairs. Leibniz buttressed his resolution with the further claim that the states the absolutist counts as different are "indistinguishable"; at least they are not separated by any features that are observable in principle. Of course, if we introduce a reference body that is not subjected to the shift, then observable differences will arise, but then the relationist will also want to count the states as distinct, since in these states there will be differences in the relations of the original system of bodies with respect to the reference body.

Many commentators assume that Kant was acquainted with the Leibniz–Clarke correspondence, and some, such as Robert Paul Wolff (1969), even assume that Kant read Leibniz's operation of "changing East to West" not as a continuous transformation (translation or rotation) but as mirror-image reflection. I know of no definitive evidence that Kant did read the Leibniz–Clarke correspondence, but it seems more likely than not, as Kant was intensely interested in Leibniz during this period and two German editions of the correspondence were available, both with forewords by Christian Wolff.[3] However, the German translation does not naturally suggest mirror-image reflection; for instance, the relevant passage in the 1740 edition reads "durch eine Verwechselung des Aufgangs der Sonnen mit ihrem Niedergangs" (literally, "through an interchange of the rising and the setting of the sun").

Moreover, if Leibniz's argument did in fact suggest the problem of incongruent counterparts to Kant, it would be surprising if Kant did not consider that the relationist can transfer Leibniz's argument as applied to continuous symmetry transformations to the case of reflections. One can speculate that Kant would have thought that the parallelism breaks down, since he would have thought that in the case of hands the situations that the absolutist wants to account as different are perceptually distinguishable: in one case the hand presents itself as a right hand, in the other as a left hand, and the difference, he may have thought, is not due to some difference in the relations of the hands to the observer. I will have more to say on this point in section 2 below.

Another response Kant may have intended can be discerned from his characterization of incongruent counterparts. "As the surface limiting the bodily space of the one cannot serve as a limit for the other, twist and turn it how one will, this difference must, therefore, be such as rests on an *inner principle*" (1768, p. 42; italics added). C. D. Broad interprets Kant as arguing that since the unlikeness between two incongruent counterparts depends upon an unlikeness between their bounding surfaces and since the surface of a body is something intrinsic to it, the difference between incongruent counterparts must rest on "a difference in their *intrinsic spatial properties* and not on a difference in *their spatial relations to some third body*" (Broad 1978, p. 38). But if this is Kant's line, then it is not at all apparent how introducing an absolute container space will help in distinguishing right from left. Suppose, in accordance with the absolute theory, that the spatial relations among the material parts of a body are parasitic upon the spatial relations of the points of absolute space occupied by the body, and consider two bodies, B and B', which are shaped like human hands and are exact mirror images of one another. Broad imagines that the set of space points occupied by B and the set of points occupied by B' might differ in some geometrical properties that are not manifested in the relations among the occupying particles of matter. But these would have to be mysterious properties indeed. If the spatial points occupied by B have the property of standing in a left-hand configuration, while those occupied by B' have the property of standing in a right-hand configuration, why can't the material points occupying the spatial points have corresponding properties?

This brings us back to (K2). If the relations among the material parts of a hand are construed narrowly to involve only, say, distance, line, and angle, then it is certainly true that the relation and situation of the parts of the hand with respect to one another are exactly similar in the cases of mirror-image left and right hands. But this is equally true if space points are substituted for material points. The absolutist may hope to distinguish between right and left by construing 'relation and situation of the space points occupied by the hand' more broadly. But it is not at all clear why the relationist cannot entertain a similar hope by appealing to a broadened notion of relation and situation of the material points of the hand.

It can be shown that, in general, being right- or left-handed cannot be purely a matter of the internal relations and situation of the material parts of a hand, no matter how broadly relation and situation are construed. Suppose, on the contrary, that it were. Then it follows that for any choice of closed path in space, it is always possible to arrange a consistent sequence of hands around the loop, where the consistency condition is that immediately adjacent hands have the same handedness. For at each location on the loop it is sufficient to construct a material hand whose parts instantiate the list of properties and relations that constitute being (say) right-handed. But nonorientable spaces show that such a consistent arrangement is not possible for every closed loop. (Here it is necessary to introduce a little technical apparatus. By 'space' I mean (in absolutist terms) something having, at a minimum, a manifold structure. The manifold may carry additional structures, e.g., affine or metric, and while such additional structures are needed for the discussion below, they are not needed for the definition of orientability. If such a space is n-dimensional, it is said to be *orientable* just in case there exists a continuous, nonvanishing field of n-ads of linearly independent tangent vectors. Equivalently, choose any closed loop in the space and erect an n-ad of linearly independent vectors at some point on the loop. Then carry the n-ad around the loop by any means of transport that is continuous and keeps the vectors linearly independent. Then upon return to the starting point the transported n-ad should not differ from the original by the reflection of any axis.)

Of course, the same argument suffices to show that being right- or left-handed cannot be purely a matter of the internal relations and situation of the points of absolute space occupied by the body.[4] Nor does it help the absolutist to bring in relations that the points of the hand have to particular points of the external space surrounding the hand, since whatever relations a left hand has to those points are exactly mirrored by the relations its right-handed counterpart has to similarly situated external points. Perhaps, however, as Nerlich (1973, 1976) suggests, the absolutist can show that the difference between right and left must make reference to the relation the hand has not just to particular points of the external space but, to use Kant's phrase, to its relations to "space in general as a unity."[5] To see the point, suppose that space is equipped with a metric that conforms to the "axiom of free mobility" so that it is meaningful to speak of the rigid transport of a body.[6]

DEFINITION An object O is an *enantiomorph* just in case there is a neighborhood N of O such that N is large enough to admit reflections of O and that the result of every reflection of O in N differs from the result of every rigid motion of O in N.[7]

The restriction to local reflections and local rigid motion is necessary if we want to be able to distinguish between handed and nonhanded objects regardless of whether space is globally orientable.[8] On the other hand, no such restriction seems justified when it comes to defining incongruent counterparts.

DEFINITION Objects O and O' are *incongruent counterparts* just in case their surfaces cannot be made to coincide by any rigid motions but can be by means of a combination of rigid motions and a reflection.

The second definition does appeal to space as a unity in that it quantifies over all motions of a certain type in the space. It has as a consequence that in nonorientable spaces (where space as a unity has a kinked structure) there are no incongruent counterparts. But that is no surprise; indeed, this consequence is just a way of restating the essential element of the definition of a nonorientable space.

What is the relevance of all of this to Kant's attempted refutation of the relational theory of space? Very little. Nerlich is incorrect in holding that Kant is right if we interpret him as saying that the enantiomorphism of a hand depends upon the relation between it and the absolute container space considered as a unity. For in the first place, my definition of enantiomorphism does not quantify over all mappings in the space. And in the second place, all that has been shown is that the absolutist has an account of enantiomorphism—something never in doubt—and not that the relationist cannot produce such an account. As for the notion of incongruent counterparts, its definition does appeal to space as a unity in Nerlich's sense. But again, Kant has not been shown to have, or to have anticipated, some insight into the right-left distinction that is beyond the ken of the relationist. Perhaps the point is that relationism is inadequate to deal with the global structure of space (orientability in particular and perhaps other global properties as well). If true, that would be a damning indictment of relationism, but it remains to be substantiated. And even if it could be substantiated, relationism would be condemned for reasons quite remote from those put forward in Kant's 1768 essay.

In closing this section I wish to emphasize the difficult position into which Kant maneuvered himself in the 1768 essay. The difference between right and left must, he asserts, rest on an inner principle. The relationist is incapable of supplying this principle, since the "difference cannot ... be connected with the different way in which the parts of the body are connected with each other " (Kant 1768, p. 42). But the absolutist can do no better in this regard, since the difference cannot be connected with the different way the points of absolute space occupied by the body are connected with each other. Rather than remarking on this obvious tension, Kant masks it by going on to assert that the "complete principle of determining physical form" rests on the relation of the object to "general absolute space" (p. 41). But if Kant's second assertion is not to be incompatible with the first, it can only mean that the inner difference, for which neither the absolutist nor the relationist can account, reveals itself through the external relation of the hand to space in general as a unity. The case of the hand standing alone shows, Kant thinks, that the relationist is blocked from claiming that the difference is revealed through the relation of the hand to other objects. But at best, this only shows that of two inadequate theories, the relational theory is more inadequate. I will return to Kant's conundrum in section 4. But first I want to turn to a more systematic examination of the possibility of a relational account of the difference between right and left.

2 A Relationist Account of the Distinction between Right and Left

As explained in chapter 6, Leibniz's criticism of Newton in the Leibniz–Clarke correspondence suggests that from the relationist perspective, the proper way to construe absolutist talk about space is to read it as providing representations of relational models of reality. These representations involve the fictional entities of space points, and consequently, the representation relation is one-many, with many absolutist representations corresponding to the same relationist model. In investigating the implications of the difference between right and left for the absolutist-relationist controversy, I propose to explore the possibility of extending this version of relationism to cover the cases at issue.

When asking whether the relationist can account for the difference between right and left, one finds it useful to distinguish between three items: (1) being an enantiomorph, (2) being incongruent counterparts, and (3) being right- or left-handed. In considering the possibility of Leibnizian accounts of (1) through (3), it is convenient to suppose that space is two-dimensional.

Being an enantiomorph

Consider the following relational description of an object. The object has a long shaft and two shorter shafts that are attached perpendicular to the same side of the long shaft. One of the cross shafts is attached to an end of the long shaft, while the second cross shaft, which is shorter than the first, is attached to the long shaft at a point between the midpoint and the end to which the first cross shaft is attached. Many different absolutist representations of this relationist description are possible, principally (*a*) and (*b*) of figure 7.1 (where the points of the page represent points of absolute space) and also rigid rotations and translations of (*a*) and (*b*). In response to Kant's example of a hand standing alone, the relationist can say that Kant was partially correct; namely, a hand standing alone is a hand: it has a handedness. More specifically, the relational hand-in-itself possesses various absolute representations viz., (*a*) and (*b*) but on any such representation, the hand is an enantiomorph, as defined in section 1 above. The relationist strategy, then, is to show that for any given object, there is a relationist description that will guarantee that any allowed absolutist representation will fulfill, or fail to fulfill, as the case may be, the absolutist definition of enantiomorph.[9]

Being incongruent counterparts

Now consider the following relational description. There are two objects, one red, the other green. Both have the characteristics described in the previous subsection. The lengths of the corresponding long and cross shafts in the two objects are the same. The objects are so situated with respect to one another that the free ends of the longer of the cross shafts are almost touching, their long shafts are parallel, and the corresponding cross shafts each lie on a line. Absolute representations consist of (*c*) and (*d*) of figure 7.2 and rigid rotations and translations of (*c*) and (*d*). Again the relational description assures that on any absolute representation, the red and green objects are incongruent counterparts, as defined in section 1. At least, this will be so if the objects are embedded in an orientable space. The question arises as to whether the orientability of the embedding space can be indicated in relational terms. It might seem that the answer is an obvious yes in the present case. For the only way a space can fail to be orientable is for it to be nonsimply connected, and such a multiple connectedness would necessarily involve a multiple relatedness of the objects. Such a multiplicity is absent in the above relational description, which we may take to be a complete catalog of the relations between the red and green objects. The hitch is that multiply connected spaces need not be nonorientable, so the absence of multiple relatedness in a relational model is a sufficient but not a necessary signal that the embedding space should be

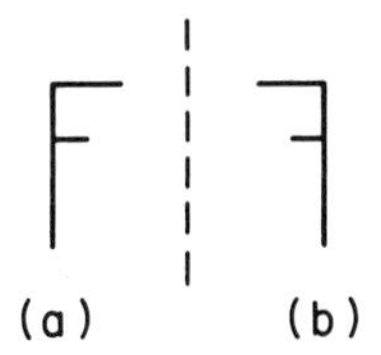

Figure 7.1
Representations of an enantiomorph

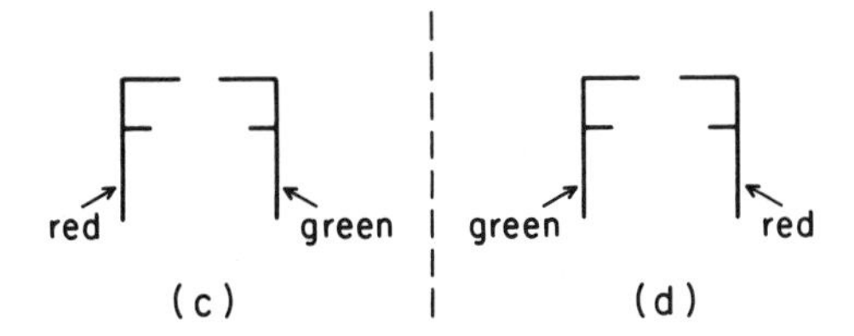

Figure 7.2
Representations of incongruent counterparts

orientable. I will not speculate here on how the relationist might respond to this difficulty.[10]

Being right- or left-handed

Here the relationist can repeat the suggestion made in section 1. The difference between right and left in the sense of Which is which? requires the choice of a reference body (say, the red one in figure 7.2) and the stipulation that it is (say) right-handed. Then being right- or left-handed is just a matter of bearing the appropriate relations to the reference standard.

A seeming difficulty with the above version of relationism (already hinted at in section 1) is that a minimal condition for counting (*a*)–(*b*) and (*c*)–(*d*) as equivalent descriptions of the same reality is that they be observationally indistinguishable . But in one sense they are patently distinguishable: they look different. The relationist will respond that talk of appearances presupposes an observer and that the introduction of an observer involves several distinct cases. If the observer is nonenantiomorphic, (*a*) and (*b*) of figure 7.1 are replaced by $(\hat{a})$ and $(\hat{b})$ of figure 7.3. If the observer is enantiomorphic, the replacements can be either $(a^{\dagger})$ and $(b^{\dagger})$ or (a^{*}) and (b^{*}). Similar replacements apply to (*c*) and (*d*) of figure 7.2. In the case of $(a^{\dagger})$ and $(b^{\dagger})$ the introduction of the observer breaks the relational equivalence of (*a*) and

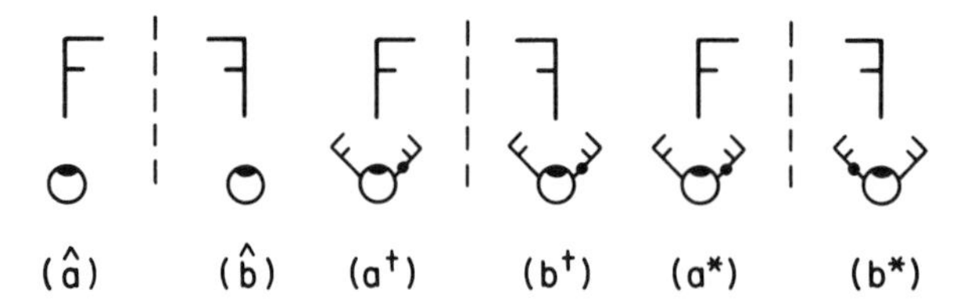

Figure 7.3
Perceptions of right and left

(*b*), and so the relationist can entertain the hypothesis that the phenomenal contents of the visual experiences of the ($a^\dagger$) and ($b^\dagger$) observers are different. Perhaps Kant would want to claim that the phenomenal contents of the experiences of the ($\hat{a}$) and ($\hat{b}$) and the (a^*) and (b^*) observers are or can be different, a claim the relationist must deny, since ($\hat{a}$)–($\hat{b}$) and (a^*)–(b^*) are relationally equivalent.

Walker might seem to be endorsing the Kantian claim when he writes, "A left-handed and a right-handed glove look different regardless of their relations to other things, and a left-handed and a right-handed universe would look different too" (1978, p. 47). Nor, according to Walker, is the difference in appearances to be accounted for in terms of the relations of the gloves to the asymmetrical body of the observer, for "the difference between the gloves is immediately obvious without reference to my body or to anything else; I should perceive it in just the same way if my body were itself symmetrical about the plane that forms the axis of symmetry between the gloves" (p. 47). But what the opponent of relationism must claim in the case of ($\hat{a}$) and ($\hat{b}$) is that the mirror-image nonenantiomorphic observers, who have the receptor sites of their sense organs left-right reversed, nevertheless have different visual experiences. Even on the absolutist's own terms, the case for this claim is weak. To support the Kantian claim, the absolutist must maintain that spatial perceptions are a function not only of the spatial relations of observer to object but also of the relation of the observer to a preferred orientation of the space containing the object-observer system. As a hypothesis about perception this seems farfetched. And postulating the preferred orientation seems ad hoc. In rejecting the relationist notion that all motion is the relative motion of bodies, the absolutist can point to the need to postulate a preferred family of reference frames, the inertial frames, in order to achieve simple but predictively accurate laws of motion. By contrast, nothing in the fundamental phenomena of physics seems to call for a preferred orientation of space. At least this was so until quite recently, as will be discussed below in section 3. But the laws in question concern exotic weak interactions of elementary particles, interactions that are presumably irrelevent to the human perceptual process.

I conclude that on the version of relationism I have proposed, the introduction of enantiomorphs and incongruent conterparts does not by itself alter the dialectics of the absolute-relational debate. Nevertheless, it might be thought that although what I called the Kantian claim does not help to establish that objects are literally embedded in absolute space, it suggests that objects as they present themselves to us in perceptions are spatial in a sense that outstrips the relational account, and thus, the claim may be useful in helping to explain how the mature Kant came to understand the implications of incongruent counterparts. Unfortunately, as will be discussed in section 4, it is no help whatever in this regard, for after 1768 Kant used incongruent counterparts not as an objection to the Leibnizian view per se but rather as a reason to reject both the Leibnizian and the Newtonian views.

3 Parity Nonconservation

In the 1768 essay Kant mentions various contingent left-right asymmetries, viz., that most hops wind round their poles from left to right, while most beans twist in the opposite direction, but nothing in his argument against relationism relies on any nonessential properties of incongruent counterparts. While this feature of Kant's argument promises to make it powerful, it may in fact account for its ineffectualness. I now propose to ask whether adding contingent but lawlike features will help.

The discussion that follows is conducted under the assumption that parity-nonconservation experiments in elementary particle physics indicate that mirror image reflection fails to be a symmetry of some of the fundamental laws of physics.[11] The experiment of Crawford et al. (1957) tracks first the decay of a negative pi meson (π^-) and a proton into a neutral hyperon (Λ°) and neutral K meson (K°), and then the subsequent decay of the hyperon into another pi meson and a proton. The momentum vectors for the initial decay process lie in a plane, while the momentum vector for the pi meson in the subsequent decay is oblique to this plane. The possible results, (*e*) and (*f*), are pictured in figure 7.4 (protons are not shown).[12] If

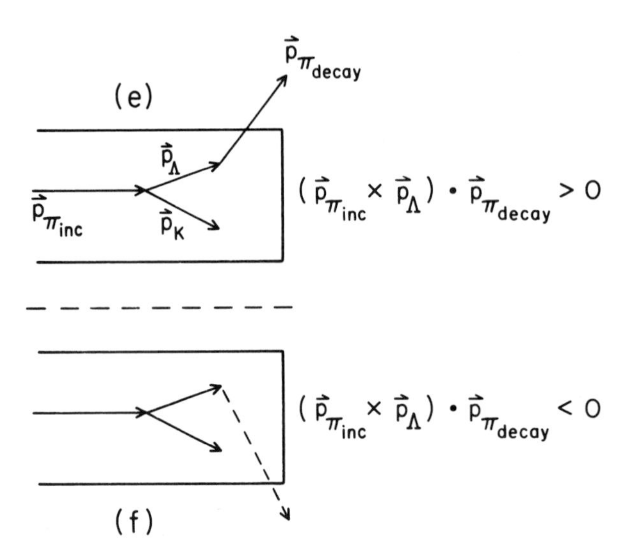

Figure 7.4
Parity nonconservation in Λ decay

parity is conserved, the mirror-image processes (*e*) and (*f*) should have the same probability, but in fact (*e*) dominates.[13] Nature's preference for (*e*) can then be used almost surely to define right-handedness:[14] perform a large number of repetitions of the decay experiment to identify the dominant decay mode, and then $\vec{p}_{\pi_{inc}}$, $\vec{p}_\Lambda$, and $\vec{p}_{\pi_{decay}}$, taken in that order, define a right-handed triad.

The failure of mirror-image reflection to be a symmetry of laws of nature is an embarrassment for the relationist account sketched in section 2, for as it stands, that account does not have the analytical resources for expressing the lawlike left-right asymmetry for the analogue of Kant's hand standing alone. If we may put some twentieth-century words into Kant's mouth, let it be imagined that the first created process is a $\pi^- + p \rightarrow \Lambda^\circ + K^\circ$, $\Lambda^\circ \rightarrow \pi^- + p$ decay. The absolutist has no problem in writing laws in which (*e*) is more probable than (*f*), but the relationist of section 2 certainly does, since for him (*e*) and (*f*) are supposed to be merely different modes of presentation of the same relational model. Evidently, to accommodate the new physics, relational models must be more variegated than initially thought.

The variegation of relationist models of particle-decay processes can proceed by the addition of intrinsic properties R^* and L^*. The relationist may not be able to describe R^* and L^* in traditional relationist terms, but he can give a functional specification of these properties in terms of their roles in the lawlike dispositions that ground the nonconservation of parity. The relationist can say that in the relevent experimental set up, R^* outcomes have a greater propensity to occur than L^* outcomes and that these propensities will almost surely be reflected in the long-run relative frequencies of multiple repetitions of the decay experiment.

There is precedent for introducing intrinsic monadic properties into relational models, Sklar's treatment of absolute acceleration being a prime example (see section 6.9). That example turned out to involve a sleight of hand, and to remove the suspicion that we are being treated to a similar performance with R^* and L^*, the relationist must tell us more about these properties. The relationist has to tell us enough about these properties to convince us that his account deserves to be called relationist in the minimal sense that R^* and L^* are not merely devices for naming absolutist states. At the same time, however, there has to be a close-enough relation between R^* and L^* on the one hand and absolutist states on the other to establish an explanatory connection between the differential propensities for R^* and L^* and the observed asymmetry between (*e*) and (*f*). The problem here has two aspects: the first and most fundamental is how and why R^* and L^* connect with (*e*) and (*f*) at all; the second is why R^* is manifested as (*e*) and L^* is manifested as (*f*) and not the other way round. The relationist may refuse to accept the second challenge and claim that there is no truth to the matter as to whether (*e*) or (*f*) is the correct representation of the dominant R^* decay mode. In this connection it is interesting to note that Lee and Yang (1956) speculated that there are actually two species of elementary particles π_R^-, Λ_R°, K_R° and π_L^-, Λ_L°, K_L°, with the corresponding particles having the same masses, charges, and spins. Parity conservation would be maintained if the two species transform into one another under mirror-image reflection and if they exhibit the opposite asymmetries in their decay modes. The apparent violation of parity conservation would be due to the fact that we inhabit a region where the R species is predominant. Although there is no experimental evidence to support this speculation, the two species do exist for the relationist as different representations or pictures of the absolutist models. That what we actually see conforms to one rather than to another of these pictures might be the basis for an objection to relationism, but it is the same objection already considered in section 2.

Now suppose for sake of argument that some suitable connection has been established between R^* and L^* on the one hand and (*e*) and (*f*) on the other. It might seem then that parity nonconservation implies that being right- or left-handed can be explicated in terms of the internal properties of a system, which contradicts the construction of section 1 involving nonorientable spaces. The conflict is only apparent. If, as I have assumed, parity nonconservation entails a violation of left-right symmetry in the basic laws of nature and if laws are universal in the sense that they hold good in all regions of space and time,[15] then it follows that actual space is in fact orientable.[16] In a hypothetical nonorientable space either the laws exhibit no left-right asymmetry, or else there are no universal laws. The absolutist may still wish to object that it remains mysterious why in nonorientable spaces the properties R^* and L^* either are not possessed or are possessed but are not manifested as right-left asymmetries. Whether or not the relationist can produce a satisfying reply depends upon two matters left hanging above: (1) the relationist's explanation of the connection between R^*, L^* and spatial representations and (2) the ability of the relationist to account for orientability properties of space.

4 Incongruent Counterparts and the Intuitive Nature of Space

The most conservative interpretation of the shifting roles Kant assigned to incongruent counterparts would locate the cause of the shift in the expansion of Kant's possibility set.[17] In 1768 the possibilities Kant considered included only Newtonian substantival space and Leibnizian relational space; after 1768 the possibility set is enlarged to include the view that space and time belong to "a form of intuition and therefore to the subjective constitution of our minds, apart from which they could not be ascribed to anything whatever" (Kant 1781, A23/B39). On the conservative interpretation, Kant used incongruent counterparts both before and after 1768 to show that space is nonrelational. Thus, to be consistent with the 1768 essay, Kant must have developed after 1768 an independent reason for rejecting the Newtonian view. Such a reason is to be found, for example, in Kant's *Inaugural Dissertation*, where the Newtonian view is denigrated this way: "The [Newtonian] empty figment of reason, since it imagines an infinity of real relations without any things which are so related, pertains to the world of fable" (Kant 1770, p. 62).

One problem with this conservative reading is that after 1768 Kant never used incongruent counterparts as an argument against Leibnizian relationism per se. In the *Inaugural Dissertation*, Kant's complaint against Leibniz and his followers is that "they dash down geometry from the supreme height of certainty, reducing it to the rank of those sciences whose principles are empirical. For if all properties of space are borrowed only from external relations through experience, geometrical axioms do not possess universality, but only that comparative universality which is acquired through induction" (Kant 1770, p. 62).[18] In both the *Inaugural Dissertation* and the *Prolegomena*, incongruent counterparts are used as an argument against relational space. Yet it is not an argument against relational space versus absolute space (as in 1768) but an argument against any conception of space, relational or absolute, that would make space something objective. Another difficulty is that after 1768 Kant used incongruent counterparts to argue directly that space is something intuited and not merely conceived (the *Inaugural Dissertation*) and that space is not a quality inherent in things in themselves (the *Prolegomena*). The argument does not have the indirect pattern required by the conservative reading.

More important, the conservative reading neglects the real possibility that incongruent counterparts were not merely recycled to buttress Kant's mature view of space but rather played a direct role in generating this view. Several commentators have remarked that the *Inaugural Dissertation* is foreshadowed in the penultimate paragraph of the 1768 essay, where Kant says that "absolute space is not an object of external sensations, but rather a fundamental concept, which makes all these sensations possible in the first place" (p. 43). What needs to be emphasized more strongly is the impetus provided by the tension in the 1768 argument (see section 1 above). Kant could hardly have failed to be aware, if only unconsciously, that Newtonian absolute space squared no better than Leibnizian relational space with his claim that the difference between right and left rests on an inner principle. The resolution called for either an abandonment of this claim or else the exhibition of a *tertium quid* to Newton and Leibniz. But Kant was unwilling to abandon the claim, and if space is regarded as the objective structure of things in themselves, there seems to be no third alternative. Thus, the 1768 essay contained both the problem and the germ of a solution. The notion that space is something that "makes all of these sensations possible" became Kant's doctrine that space is a form of outer intuition, allowing an escape from the corner into which he had painted himself in the 1768 essay and allowing incongruent counterparts to be seen, as they were from 1770 onward, as an objection to both the absolute and relational views insofar as they deny that space belongs to the subjective constitution of the mind.

This reading of Kant makes more natural the otherwise startling and abrupt shift that occurs between 1768 and 1770. Unfortunately, confirmation is to be found not so much in the *Inaugural Dissertation* as the *Prolegomena*. In the latter, Kant repeated his original claim that the differences between right and left are "internal differences."[19] But instead of using this claim as a refutation of relationism, Kant combined it with the further claim that the differences are ones that "our understanding cannot show to be internal," and presumably this is so whether the understanding uses the concepts of Leibnizian or Newtonian theory. The upshot is that "these objects are not representations of things as they are in themselves, and as some pure understanding would cognize them, but sensuous intuitions, that is, appearances whose possibility rests upon the relation of certain things unknown in themselves to something else, viz., to our sensibility" (Kant 1783, p. 30). Thus, incongruent counterparts are being used not to adjudicate between relational and absolute space, but to reject both insofar as they are supposed to apply to things in themselves. Substituting space as the form of external intuition for Newtonian absolute space allowed Kant to retain the 1768 idea that "region is related to space in general as a unity, of which each extension must be regarded as a part." Thus, in the *Prolegomena* we find: "Space is the form of the external intuition of this sensibility, and the internal determination of any space is possible only by the determination of its external relation to the whole of space, of which it is a part (in other words, by its relation to external sense)" (Kant 1783, p. 30).[20]

If this less conservative reading of Kant's philosophical development is accurate and if the analysis of the difference between right and left given in the preceding sections is correct, we have another example of a major philosophical system that is rooted in a mare's nest of confusions.

5 Conclusion

Whatever the correct account of the role of incongruent counterparts in Kant's philosophical development, it is hard to find in his writings the suggestion for an even half-way-plausible argument to the effect that the very existence of incongruent counterparts establishes the reality of absolute space. Adding the consideration of parity nonconservation makes an interesting but not decisive difference. If he is willing to add enough epicycles to his theory, the relationist can deflect any objection launched by the absolutist. The need to add epicycles is not necessarily an indication of falsity, but the accumulation of enough epicycles may cause one to lose interest in the theory.[21]

8 Modern Treatments of Substantivalism and Relationism

Chapters 6 and 7 reviewed two arguments in favor of substantivalism: the argument from absolute motion and the argument from incongruent counterparts. The first argument contends that relationist thesis (R1) fails and that this failure carries with it a failure of (R2). The second claims that a proper understanding of the right-left distinction entails that (R2) is false. In this chapter I shall review a third argument for substantivalism, one championed by Hartry Field (1980, 1985). Although akin to the first argument, it is independent in that it purports to show that even if the structure of space-time were no stronger than the relationist would have it, (R2) fails, because it cannot adequately provide for the field theory of modern physics.

A possible counter to all these arguments is the representationalist ploy explored in chapters 6 and 7. In sections 5 and 6, I shall review some recent technical elaborations of this ploy. It has been claimed by some (e.g., Friedman 1983) that these elaborations clarify the issues and by others (e.g., Mundy 1983) that they show how relationism is a viable and even correct view. I shall register some reasons for being skeptical about these claims. What I do find worth exploring, however, is the possibility that if exploited along the lines suggested by Sklar (see section 6.9), the representationalist ploy may lead to a *tertium quid* lying between the traditional relational and absolute conceptions of space and time.

1 Field's View

In *Science without Numbers* Hartry Field offers two reasons for a substantivalist interpretation of space-time. The first is that it supports his nominalist goals: being a substantivalist with respect to space-time allows one to quantify over space-time points with a good nominalist conscience, and such quantification dovetails with Field's program of showing how to do science without resort to the platonic entities nominalists find objectionable. I take no stand for or against nominalism here and thus shall ignore Field's first motivation.

Field's second reason is both less self-serving and more profound.

I don't think that any relationist program, of either a reductive or an eliminative sort, has ever been satisfactorily carried out....[1] The problem for relationism is *especially* acute in the context of theories that take the notion of *field* seriously, e.g., classical electromagnetic theory. From the platonistic point of view, a field is usually described as an assignment of some property, or some number or vector or tensor, to each point of space-time; obviously this assumes that there are space-time points, so a relationist is going to have to either avoid postulating fields (a hard road to take in modern physics, I believe) or else come up with some very different way of describing them. The only alternative way of describing fields that I know is the one I use later in the monograph.... It does without the properties or numbers or vectors or tensors, but it does not do without space-time points. (Field 1980, p. 35)

Field's remarks here may be construed as endorsing what I have called manifold substantivalism. Departing slightly from what he says about classical electromagnetism, I would rephrase his point as follows. When relativity theory banished the ether, the space-time manifold M began to function as a kind of dematerialized ether needed to support the fields. In the nineteenth century the electromagnetic field was construed as the state of a material medium, the luminiferous ether; in postrelativity theory it seems that the electromagnetic field, and indeed all physical fields, must be construed as states of M. In a modern, pure field-theoretic physics, M functions as the basic substance, that is, the basic object of predication. These ideas will be explored in more detail in section 3. But first I want to examine the folklore wisdom that relativity theory requires the presence of fields because it is inconsistent with particles interacting at a distance.

2 Relativity and Fields

Is relativity theory inconsistent with the view that particles can act on one another at a distance without mediating fields? The question is partly terminological; it turns on what is counted as part of the theory of relativity and what is counted as a field. But it also engages substantive issues about the nature of scientific theorizing and the form and content of physical theories.

STR is not a theory in the usual sense but is better regarded as a second-level theory, or a theory of theories that constrains first-level theories. The main constraint is Lorentz (or more properly, Poincaré) invariance. In terms of the structure of space-time models, the demand is that the theory be formulated in the arena of Minkowski space-time, the understanding being that the theory not smuggle in any additional structure for space-time, such as distinguished reference frames. As the remarks in the appendix to chapter 2 were supposed to indicate, it is a delicate and difficult task to separate the object fields into those that characterize the space-time structure and those that characterize its physical contents. But the vagaries of this general problem need not detain us here, since there are enough clear cases for our purposes.

If Lorentz invariance were the only constraint that STR imposes, then it is clear that STR per se does not demand the mediation of fields in particle interactions. The physics literature contains a large number of nontrivial, Lorentz-invariant, pure particle theories; the interparticle interactions are variously conceived as taking place along the retarded and/or advanced light cones (as in the Wheeler–Feynman theory), along spacelike intervals (as in the van Dam and Wigner theories), or instantaneously at a distance (as in the theories of Currie and Hill).[2]

It remains open, of course, that Lorentz invariance and some other requirements that one would expect an adequate physical theory to satisfy together entail the need for fields as mediating agents. The additional requirements that come most readily to mind are conservation principles. In Newtonian theories, energy might be attributed to a field postulated to mediate the particle interactions, but it is otiose to attribute momentum to the field if, as is usually the case, the total momentum of the particles is conserved. Relativity theory turns the tables, since in relativity theory it is the combination of energy-momentum that is conserved, if at all, and if influences propagate with a finite speed, it would seem that the energy-momentum of particles cannot be conserved without the help of a field that carries some of it. This intuition is made precise in a simple and ingenious theorem due to van Dam and Wigner (1966). This theorem shows that no nontrivial, pure particle theory can satisfy both Lorentz invariance and a conservation law for energy-momentum of the form

$$P^i = \text{constant for } i = 1, 2, 3, 4, \tag{8.1}$$

where P^i is the total linear energy-momentum of the system of particles. More precisely, if $P^i = \sum_k^N P_k^i$ (the sum being taken over the individual particles labeled by k) is independent of t in any Lorentz frame (x, t), the particles do not collide, and asymptotically (as the proper time $\tau_k \to \pm\infty$) the particle orbits are straight lines, then for $N = 2, 3$, or 4 the orbits are straight lines for all time.

Van Dam and Wigner show that in their theory of action-at-a-distance particle mechanics the conservation law can be restored in the form

$$P^i + V^i = \text{constant}, \tag{8.2}$$

where V^i is the interaction momentum of the system. But since V^i involves an integral over the actual orbits of the particles, (8.2) is seen by some commentators as a mathematical trick, for the integration involved is "precisely what we would regard as field momentum in a field theory (the integration over history arises from writing the field in terms of the past motions of all the particles that contributed to the field)" (Ohanian 1976, p. 80).

In response the particle theorist may say that we simply have to live with the fact that conservation laws in the familiar form cannot hold for each instant of time but only asymptotically for particles that are widely separated both initially and finally, in which case we may be able to prove that incoming energy-momentum equals outgoing energy-momentum. Second, while the introduction of fields as storehouses of energy and momentum may facilitate the maintenance of cherished forms of conservation principles, it may also give rise to other problems. For instance, infinities can arise from particles conceived as creating fields, which in turn act back on the particles. The infinities can sometimes be suppressed by means of clever subtraction procedures, but these procedures are no less artificial than the introduction of the interaction momentum in (8.2). It was precisely the avoidance of such infinities that motivated some of the action-at-a-distance formulations of classical relativistic electrodynamics. At least those theories have the virtue of consistency.

Nevertheless, it is hard to shake the feeling that in the context of STR, pure particle theories are artificial, and that various features of such theories indicate that fields are struggling to emerge from the formalism. For instance, in the Currie–Hill theories of instantaneous action at a distance it is found that in multiparticle systems the total force acting on a particle is not the sum of two-body forces, and the presence of multibody forces can be read as the vestige of the fields that were suppressed in the pure particle description (see Hill 1967a). Furthermore, in these theories it is apparently inconsistent to treat the measurement of particle position, say, as a purely local interaction between the particle and the measuring apparatus. Hill (1967b) proposed to restore consistency by postulating a nonlocal interaction between the measuring apparatus and the other particles. The postulate remains a bare postulate.

More straightforward considerations of empirical adequacy can be brought to bear against certain action-at-a-distance theories. Poincaré-type gravitational theories, where the particles interact along retarded light cones, can be rigged to yield a correct value for the advance of the perihelion of Mercury. But there is no natural way to use such theories to explain both the red shift and the bending of light (see Whitrow and Morduch 1965).

In sum, there are a variety of reasons to be uneasy about special-relativistic theories that attempt to do without fields. The reasons are a mixture of considerations of empirical adequacy, the form of conservation laws, naturalness, etc. These considerations may be persuasive to one degree or another, but they do not add up to an irrefutable proof that STR demands that interactions be mediated by fields. But then it may be a mistake to look for proofs as opposed to persuasions on such questions. When the persuasions of relativistic quantum field theory and the general-relativistic theory of gravitation are added, the case becomes very persuasive indeed.

3 Fields and Manifold Substantivalism

It is useful to flesh out Field's conception of space-time as a dematerialized ether. To do this, we need to review the standard definition of an n-dimensional manifold.[3] Intuitively, M is made up by piecing together in a smooth way open sets of $\mathbb{R}^n$. To make this precise, we specify that M consists of a set of points together with an atlas of charts $\{U_\alpha, \psi_\alpha\}$, where the U_α are subsets and the ψ_α are one-to-one mappings from the corresponding U_α to open sets of $\mathbb{R}^n$ such that (1) $\bigcup_\alpha U_\alpha = M$, and (2) if $U_\alpha \cap U_\beta \neq \varnothing$, then $\psi_\alpha \circ \psi_\beta^{-1} : \psi_\beta(U_\alpha \cap U_\beta) \to \psi_\alpha(U_\alpha \cap U_\beta)$ is a smooth map of an open set of $\mathbb{R}^n$ to an open set of $\mathbb{R}^n$. Another atlas is said to be compatible with the given one if their union is again an atlas. We assume that the atlas we are working with is complete in that it contains all of the atlases compatible with the given one. The topology of M is then fixed by stating that the open sets of M are unions of the U_α belonging to the complete atlas. Alternatively, we could have started with a topological space and required that the ψ_α are homomorphisms. Physicists refer to the ψ_α as coordinate systems and use the notation x^i (i.e., $\psi_\alpha(p) = (x^1(p), x^2(p), \ldots, x^n(p))$ for $p \in M$), a convention already adopted in the chapters above. In space-time theories, M is usually assumed to be Hausdorff, paracompact, and without boundary, an assumption that will be in force in what follows unless otherwise specified.[4]

A geometric object field on M is a correspondence $F : (p, x^i) \to (F_1, F_2, \ldots, F_N) \in \mathbb{R}^N$ that assigns to each $p \in M$ and each coordinate system x^i about p an N-tuple of real numbers, called the coordinate components at

p of F in the coordinate system x^i, and that satisfies the condition that the new components F'_K given by $F:(p, x'^i) \rightarrow (F'_1, F'_2, \ldots, F'_N)$ in the new coordinates x'^i are determined as functions of the old components F_J and the coordinate transformation $x'^k = f(x^j)$.[5] Although this definition appeals to coordinate systems and transformation of components under coordinate changes, it is designed to pick out a set of objects that deserve to be called 'invariants' in the sense that the coordinate components of the object are aptly named: they are components of an intrinsic object that lies behind all of the many different coordinate representations. The most familiar geometric objects—vectors and tensors—are linear and thus can be given direct, coordinate-free characterizations. For example, a tangent vector V at $p \in M$ is a derivation of the smooth real valued functions $\mathfrak{F}$ on M, i.e., V is a map from $\mathfrak{F}$ to $\mathbb{R}$ such that (1) $V(\lambda f + \mu g) = \lambda V(f) + \mu V(g)$ for all λ, $\mu \in \mathbb{R}$ and all f, $g \in \mathfrak{F}$, and (2) $V(fg) = f(p)V(g) + g(p)V(f)$. From this definition it follows that the transformation rule for vector components is $V'^i = \sum_j V_j \, \partial x'^i / \partial x^j$, which satisfies the requirements for a geometric object. A tangent vector field is an assignment of a tangent vector V_p at each point $p \in M$; the field is smooth if for any $f \in \mathfrak{F}$, $V(f)$ is smooth. Various attempts have been made in the mathematics literature to give a coordinate-free characterization of geometric objects in general.[6] These attempts are too technical to be reviewed here.

It is clear that the standard characterization of fields uses the full manifold structure: the points, the topology, and the differentiable structure. The antisubstantivalist can, of course, attempt to dispense with some or all of this apparatus in favor of another means of specifying a relationally pure state of affairs and view all the above as merely giving representations of the underlying relational state in much the same way that the coordinate components of a vector are merely representations of an intrinsic object. While not prejudging the success of such an endeavor, I will say that the burden of proof rests with the antisubstantivalist. Before turning to a consideration of how this burden might be discharged, I want to emphasize how standard GTR depends upon M to support the metric and other fields.

4 The First Hole Construction

That M does indeed function as a kind of dematerialized ether in GTR can be brought out by means of what I shall call the holing operation. Just as

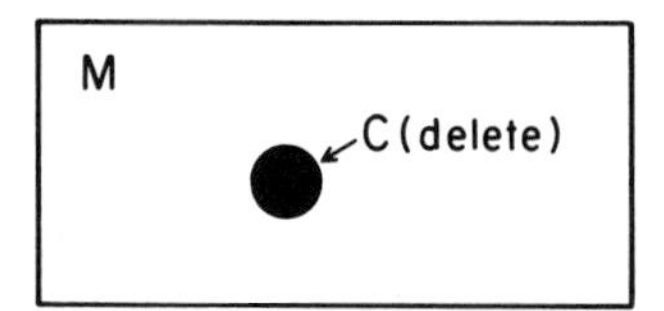

Figure 8.1
First hole construction

we can imagine that God could create a hole in the material ether, so we can imagine that He can create a hole in M. Let M, g be a relativistic space-time, and let C be a closed subset of M.[7] Surgically remove C from M to form $M^- = M - C$ (figure 8.1), and then restrict g to M^-. Note that if $\mathscr{M} = \langle M, g, T \rangle$ satisfies Einstein's field equations, then so does the holed out $\mathscr{M}^- = \langle M^-, g|_{M^-}, T|_{M^-} \rangle$. And more generally, if the models of a theory are of the form $\langle M, O_1, O_2, \ldots \rangle$, where the O_i are geometric object fields on M and the laws of the theory are in the form of local partial differential equations for the O_i, then $\langle M^-, O_1|_{M^-}, O_2|_{M^-}, \ldots \rangle$ will be dynamically possible whenever the original model was.

The substantivalist will say that in the situation specified by the holed model $\mathscr{M}^-$, no events occur or can occur in C, or more properly, since the points in C no longer exist after the holing operation, he will say that no events occur or can occur beyond certain limits taken in M^-. The relationist can tolerate some such talk if the missing points are singular—a point of view already codified in the very definition of a relativistic space-time M, g requiring that g be defined on all of M. Thus, if C corresponded to a curvature singularity in the sense that (say) the scalar curvature "blows up" along suitable curves approaching C, making it impossible to continuously extend $g|_{M^-}$, there would be a perfectly acceptable explanation for the hole in the fabric of events. But the relationist cannot tolerate talk about holes in the fabric of events if the missing points are regular, for to do so is tantamount to admitting the essence of space-time substantivalism: the notion that events are happenings at space-time points construed as ontologically prior to the happenings.

The relationist is thus faced with a dual challenge: first, he must provide a criterion to detect what the substantivalist would call missing regular points, and second, he must provide a non-question-begging reason for excluding such models from consideration.[8] At first glance the first part of the challenge does not appear to be too daunting. Let us say that M, g has

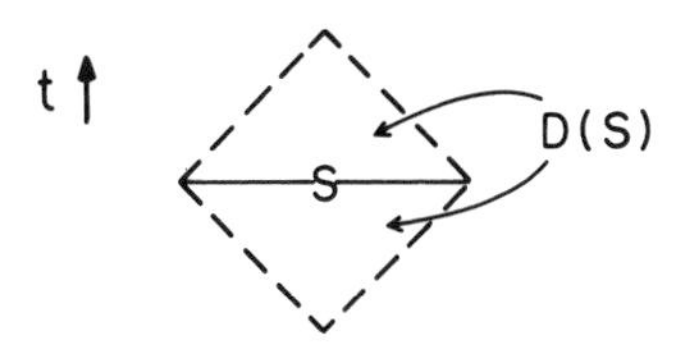

Figure 8.2
A space-time free of D-holes but not S-holes

an *S-hole* ('*S*' for 'surgical') just in case there is a space-time M', g' and an isometric embedding $\phi: M \rightarrow M'$ such that $\phi(M)$ is a proper subset of M'. This definition has been constructed to detect when the space-time in question has been subjected to a holing operation. A justification for ignoring space-times that are not *S*-hole free can be given in two steps. First, it can be shown that any space-time can be extended to a space-time that is maximal or *S*-hole free. Second, one can argue on PSR grounds that there is no good reason for the Creative Force to stop building until the maximal extent is reached, and on grounds of plenitude that the maximal model is better than a truncated submodel.

Even if one resonates to this Leibnizian line, there is more to the tune. For M, g may be *S*-hole free and yet be holey in other ways. If $S \subset M$ is a spacelike hypersurface, define the domain of dependence $D(S)$ of S to be the set of all $p \in M$ such that every smooth causal curve that meets p also meets S. Intuitively, p's being in $D(S)$ is a necessary condition for the state at S to uniquely determine the state at p, since if $p \notin D(S)$, there are possible causal influences that can affect events at p without registering on S. Say that M, g has a *D-hole* ('*D*' for 'determinism') just in case there is a spacelike $S \subset M$, a space-time M', g', and an isometric embedding $\phi: D(S) \rightarrow M'$ such that $\phi(D(S))$ is a proper subset of $D(\phi(S))$.[9] To rule out space-times that suffer from *D*-holes, one could invoke the causal version of the PSR and point out that a *D*-hole implies a spontaneous breakdown of a deterministic evolution for which there is no good reason.

Being *S*-hole free and *D*-hole free are independent requirements. The space-time in figure 8.2, created from two-dimensional Minkowski space-time by deleting all the points outside of $D(S)$, is *D*-hole free but not *S*-hole free. The result of removing a spacelike two-plane from four-dimensional Minkowski space-time and then taking the universal covering space is *S*-hole free but not *D*-hole free (see Clarke 1976). Moreover, the motivations for requiring *S*- and *D*-hole freeness can be at odds if the space-time

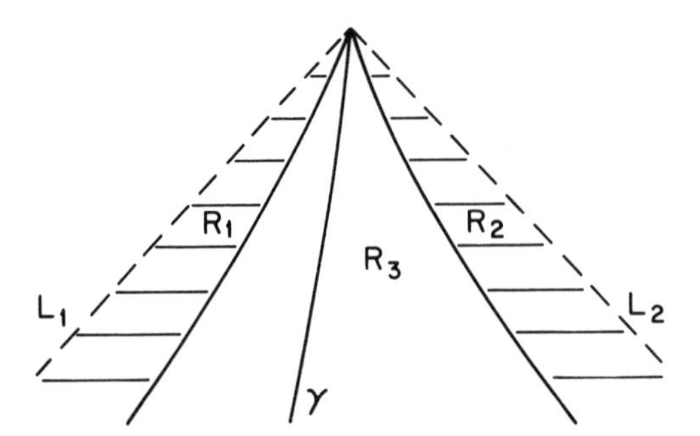

Figure 8.3
A globally inextendible but locally extendible space-time

contains latent D-holes. M, g may fail to be maximal, so by the motivation for S-hole freeness the Creative Force should go on extending M, g, but any such extension may turn up D-holes. Clarke (1976) produced an example of a D-hole-free space-time that cannot be extended to a D-hole-free maximal space-time.

Another approach to hole freeness was suggested by Hawking and Ellis (1973). Let us say that M, g has an *L-hole* ('L' for local) just in case there is an open $U \subset M$ with noncompact closure in M, a space-time M', g', and an isometric embedding $\phi : U \to M'$ such that $\phi(U)$ has compact closure in M'. Beem (1980) showed that this definition is unsatisfactory in that standard Minkowski space-time is not L-hole free. A suitable modification of the definition can be made with the help of the concept of generalized affine length discussed by Hawking and Ellis (1973, p. 259). Let us instead say that M, g has an *L-hole* just in case there is an inextendible half-curve $\gamma(\lambda)$, $\lambda \in [0, r)$, $r > 0$, of finite generalized affine length, a neighborhood U of $\gamma([0, r))$, a space-time M', g' and an isometric embedding $\phi : U \to M'$ such that the image curve $\phi(\gamma(\lambda))$ can be continuously extended through $\lambda = r$.

L-hole freeness entails both S- and D-hole freeness, but the converse does not hold, as is illustrated by an example due to Robert Geroch (private communication; see Ellis and Schmidt 1977 for other examples). Start with two-dimensional Minkowski space-time, and introduce a new metric $g = \Omega\eta$, where η is the Minkowski metric and Ω is a conformal factor. Choose Ω to be unity on region R_3 (see figure 8.3), and on the shaded regions R_1 and R_2 design Ω to blow up rapidly as the null lines L_1 and L_2 are approached from below. The resulting space-time M^*, g, where M^* consists of the union of the regions R_1, R_2, and R_3, has no S- or D-holes, but it is locally extendible. In cases such as these the relationist's appeals to principles of sufficient reason and plenitude are unavailing, but two strategies are still open to him. He can try to show that such cases can be ignored because they do not arise in physically significant situations. Failing that, he can work through the details of the case to show why the missing regular points, as detected by the existence of a local extension, cannot simply be added on to the original space-time to make a larger space-time, and he can then try to translate this reasoning into his vernacular to produce a relationist explanation of the impossibility in question. The prospects of success cannot be assessed until we know more about how the relationist proposes to make good on what is up to now only an empty boast of 'I can do all you can do, only better'.

The curious blend of precise mathematical questions and fuzzy and wild metaphysics encountered in this section is not untypical of the absolute-relational controversy.

5 Friedman on Relationism: Model-Submodel versus Model Embedding

Chapters 2 to 5 made clear, I trust, what is at stake in the absolute-relational debate over the nature of motion and the structure of space-time. And it is also apparent, I hope, why the outcome of that debate merits the interest of philosophers and physicists alike. But despite the efforts of chapters 6 and 7 and sections 1 to 4 of this chapter, I confess that one may be left wondering why anyone, other than a few academic philosophers with a bent for far-side metaphysics, should care about the tug of war over whether space-time is a substance. That concern will be addressed in chapter 9. Here I turn to further efforts to clarify what the tug of war is about.

In *Foundations of Space-Time Theories* Friedman identifies Leibnizian relationism, by which I take him to mean antisubstantivalism, with the view that the domain over which the quantifiers range is to be limited to the set of concrete physical events, "that is," he adds, "the set of space-time points that are actually occupied by material objects or processes" (1983, p. 217). The explanatory clause adds misdirection rather than clarification, or so I will argue.

Consider the illustration offered. Think of Minkowski space-time and think of absolutist models in this setting as having the form $\langle M, \tau, \sigma, \lambda \rangle$ where $M = \mathbb{R}^4$ and τ, σ, and λ are binary relations on M having the following interpretations: for $p, q \in M$, $\tau(p,q)$ (respectively, $\sigma(p,q)$, $\lambda(p,q)$) holds just in case p and q have a timelike (respectively, spacelike, lightlike) separation. A relationist model has the form $\langle \mathscr{E}, \tau', \sigma', \lambda' \rangle$, where $\mathscr{E}$ is a set of physical events and τ', σ', and λ' are binary relations on $\mathscr{E}$. In keeping with the above quoted statement of relationism, Friedman characterizes the dispute between the relationist and the substantivalist by taking the latter to hold that $\langle \mathscr{E}, \tau', \sigma', \lambda' \rangle$ is literally a submodel of $\langle M, \tau, \sigma, \lambda \rangle$, i.e., $\mathscr{E} \subseteq M$, and that the primed relations are restrictions of the unprimed relations to $\mathscr{E}$, whereas the former says only that $\langle \mathscr{E}, \tau', \sigma', \lambda' \rangle$ is embeddible into $\langle M, \tau, \sigma, \lambda \rangle$, i.e., there is a one-to-one map $\phi : \mathscr{E} \to M$ such that $\phi(\tau') = \tau|_{\phi(\mathscr{E})}$ etc. So the absolutist regards the larger structure as a reduction or explanation of the smaller structure, while the relationist regards the larger structure as a representation of the smaller one (Friedman 1983, p. 220).

A small but important caveat has to be entered at this juncture. Neither the relationist nor the substantivalist will want to say that $\mathscr{E}$ is literally a subset of M. It is a category mistake to overlap $\mathscr{E}$ and M: the former is a set of events, the latter is a set of space-time points. It is true that in the relativity literature the term 'event' is often used ambiguously to denote both events proper, i.e., happenings, and event locations or space-time points, but that ambiguity becomes pernicious when introduced into the present discussion. Thus, the relationist and substantivalist should agree that the absolutist models should be expanded to include a set E of absolutist events conceived as goings-on at space-time points, e.g., the oscillation of the magnetic field at $p \in M$. The absolutist will claim that there is a preferred embedding $\phi : \mathscr{E} \to E$, namely, the one such that for every $e \in \mathscr{E}$, e is in fact constituted by $\phi(e)$ (which for present purposes means that e occurs at the space-time location of $\phi(e)$). By contrast, the relationist will claim that for present purposes any association $\phi : \mathscr{E} \to E$ will do, as long as for any $e_1, e_2 \in \mathscr{E}$, $\tau'(e_1, e_2)$ if and only if $\tau(\phi(e_1), \phi(e_2))$, etc.

These rather pedantic distinctions do have an important consequence under the assumption that there are no unoccupied space-time points. According to Friedman's initial characterization, this assumption "means that the relationist's ontology is just as rich as the absolutist's: $\mathscr{E}$ is not just embeddible into M, it is actually isomorphic to M. Hence, the traditional debate threatens to dissolve completely" (1983, p. 222). And this worry leads Friedman to worry about how to properly characterize unoccupied points, especially in the context of field theories where the dissolving assumption might seem to hold.[10] But in our emended formulation there is no need to worry about a collapse of the debate. Even if the world were chocked completely full, the relationist's ontology would *not* be just as rich as the substantivalist's; $\mathscr{E}$ and E would be isomorphic, but the relationist would still regard M as nothing more than a construction out of his $\mathscr{E}$. And even if the world were chocked full, Leibniz's PSR argument would still apply.[11]

It is instructive to see how Leibniz's argument fares for Friedman's example. Let h be an automorphism of $\langle M, E, \tau, \sigma, \lambda \rangle$; that is, $h = (h_1, h_2)$, where $h_1 : M \to M$ is one-to-one and onto and preserves τ, σ, and λ, and where $h_2 : E \to E$ is also one-to-one and onto and such that for each $e \in E$, $h_1(\text{loc}(e)) = \text{loc}(h_2(e))$, where loc($e$) stands for the space-time location of e. Then if $\phi : \mathscr{E} \to E$ is an acceptable embedding of $\langle \mathscr{E}, \tau', \sigma', \lambda' \rangle$ into $\langle M, E, \tau, \sigma, \lambda \rangle$, then $\phi' = h_2 \circ \phi$ is an acceptable embedding into $\langle M, h_1(E), \tau, \sigma, \lambda \rangle$, where $h_1(E)$ indicates the set of events obtained by relocating the E in M under the action of h_1. The substantivalist is thus committed to distinct states of affairs $\langle M, E, \tau, \sigma, \lambda \rangle$ and $\langle M, h_1(E), \tau, \sigma, \lambda \rangle$ that are observationally indistinguishable; the relationist escapes this embarrassment, since for him the substantivalist models are only different modes of presentation of the same relationist state of affairs.

Now suppose that E, τ, σ, and λ are such that $\langle M, E, \tau, \sigma, \lambda \rangle$ admits no nontrivial automorphisms. It has seemed to many commentators that this supposition blocks Leibniz's PSR objection,[12] and since post-GTR space-time models may conform to the supposition, it has further seemed that Leibniz's concern has been left behind by modern developments. This is a myopic point of view, for h_1 may fail to preserve τ, σ, and λ, but $\phi' = h_2 \circ \phi$ is an acceptable embedding into $\langle M, h_1(E), h_1(\tau), h_1(\sigma), h_1(\lambda) \rangle$, which (as long as h_1 preserves the structure of M) is also a legitimate substantivalist state of affairs distinct but observationally indistinguishable from $\langle M, E, \tau, \sigma, \lambda \rangle$.[13]

My main objection to this entire line of discussion goes beyond the pedantic points made above. The simplicity of the examples, while allowing various structural features to stand out, conveys the misleading impression that the antisubstantivalist has staked out a viable position and that the debate boils down to a metaphysical choice between submodel and embedding. In previous chapters I tried to show why the constant appearance in the philosophical literature of the phrase "relational theory" is misleading with regard to the issue of the nature of motion. Not a single relational theory of classical motion worthy both of the name 'theory' and of serious consideration was constructed until the work of Barbour and

Bertotti in the 1960s and 1970s. This work came over half a century after classical space-time gave way to relativistic space-time, and in the latter setting a purely relational theory of motion is impossible, or so I have argued in chapter 5. Similarly, on the issue of substantivalism no detailed antisubstantivalist alternative has ever been offered in place of the field theoretic viewpoint taken in modern physics.

On the issue of substantivalism the relationist can follow either of two broad courses. One, he can decline to provide a constructive alternative field theory and instead take over all of the predictions of field theory for whatever set of quantities he regards as relationally pure. I do not see how this course is any different from instrumentalism. While I believe instrumentalism to be badly flawed, I do not intend to argue that here. Rather, the point is that relationism loses its pungency as a distinctive doctrine about the nature of space and time if it turns out to be nothing but a corollary of a methodological doctrine about the interpretation of scientific theories.

Two, the relationist can attempt to provide a constructive alternative to field theory or at least to the substantivalist version of it discussed above. This course in turn has two branches. First, the relationist could try to dispense with fields altogether in favor of a pure particle ontology. While I indicated in section 3 that this branch is not excluded by relativity theory, it is sufficiently unattractive that it can be set aside as a desperate ploy. Second, the relationist could attempt to describe physical states in a vocabulary that is sparse enough to avoid talk that smacks of space-time substantivalism but rich enough that the relational state of affairs determines the substantivalist state up to an appropriate degree of uniqueness. In section 7 I shall discuss the type of representation theorem the relationist would have to prove to make his case. But first I want to register some demurrers to the impression, promoted by Mundy (1983), that without further ado relationism is a viable alternative.

6 Mundy on Relational Theories of Euclidean Space and Minkowski Space-Time

Mundy (1983) has offered what he calls relational theories of Euclidean space and Minkowski space-time. The merits of such theories have to be judged in the larger context of relational theories of physics in general, but even in their own terms these theories are beset with two uniqueness problems.

For simplicity, consider first the case of three-space. Suppose that there are a finite number of point particles and that the mutual spatial relations among the particles obey the axioms of Euclidean geometry. For the relationist, any talk about space per se is to be analyzed as talk about representations of the interparticle relations. There are an innumerable number of ways to effect such a representation, e.g., an innumerable number of different embeddings into standard Euclidean space $\mathbb{E}^3$, and this fact is the source of Leibniz's PSR and PIdIn objections. But the relational structure in question can also be represented by embedding it into a space that is metrically Euclidean inside some finite neighborhood containing the particles but metrically non-Euclidean outside, by embedding it into a space that is obtained from $\mathbb{E}^3$ by a holing operation (see section 4 above), etc.

The relationist can react to the nonuniqueness of the embedding space with either a conventionalist or a nonconventionalist attitude. The conventionalist is prepared to say that to the extent that the embedding space is underdetermined, so is the structure of space. But many philosophers of science would choke at having to swallow the result that there is no truth to the matter of whether space is, say, infinite or finite in extent. The nonconventionalist relationist can resort to various ploys, the most obvious of which was already suggested by Leibniz's dictum "Space denotes, in terms of possibility, an order of things which exist at the same time" (Alexander 1984, p. 26). The appeal to *possibilia* can involve different grades of modal commitment, the highest degree of which would permit the relationist to gloss 'Space is infinite' as

I For any body b and any $d > 0$, there is a possible body b' such that $\text{dist}(b, b') > d$.

But it is not clear how talk about possible bodies is all that different than talk about space points as permanent possibilities of spatial location, an attitude that is arguably consistent with Newton's explanation of the status of space in "De gravitatione" (see chapters 1, 3, and 4). Descending to a lower grade of modal involvement, one could replace (I) by

I′ For any body b and any $d > 0$, it is possible that there is a body b' such that $\text{dist}(b, b') > d$.

Now the worry is that whatever reading the modal operator is given, (I′) will be either too strong or too weak.[14] If possibility in (I′) is taken to be physical possibility, then space may be infinite even though (I′) fails because, say, a repulsive force prevents particles from existing too far out toward spatial infinity. If possibility is degraded to mere logical possibility, the worry becomes that space might be finite even though (I′) holds. If 'it is possible that ———' is read as 'there is a possible world in which ———', then (I′) can be true even though space in our world is finite. This fault can be eliminated if the accessibility relation on possible worlds is limited so that worlds accessible from our world must have the same spatial structure as our world. Such a limitation might be reasonable if the relevant sense of possibility were physical possibility and the structure of space immutable, but this interpretation reengages the first worry. And in any case, specifying when two worlds have the same spatial structure seems no less difficult a task for the relationist than the original task.

This problem of uniqueness would not arise in a plenum. Toward that end, the relationist could switch the focus of attention from space to space-time, replace the particle ontology with an ontology of events, and appeal to field theory to supply a plenum of physical events. But while a plenum of events is a not unnatural circumstance in special relativistic field theories, these theories permit models in which all of the physical fields vanish over finite or infinite stretches of space-time. The g-field in GTR never vanishes, as follows from the very definition of a general relativistic space-time. But for the relationist to try to construct space-time out of events that thus far have received only a substantivalist interpretation (e.g., the events of the g-field taking such and such a value at such and such points of M) is to invite the charge of circularity.

A second form of nonuniqueness worries Mundy (1986), whose illustrative example I follow here. Suppose that we ignore the above uniqueness problem and fix upon Minkowski space-time as the embedding space-time. And suppose, in the notation of section 5, that $\mathscr{E} = \{e_1, e_2, e_3\}$ and that $\tau'(e_1, e_2)$, $\tau'(e_2, e_3)$, and $\tau'(e_1, e_3)$. Two types of embeddings or representations are possible, as illustrated in figure 8.4, namely, e_1, e_2, e_3 on a straight line in space-time (*a*) versus e_1, e_2, e_3 not on a straight line (*b*). The substantivalist will want to say that since events are literally in space-time, only one of (*a*) and (*b*) can be correct; there is truth to the matter of whether e_1, e_2, e_3 are collinear or not.

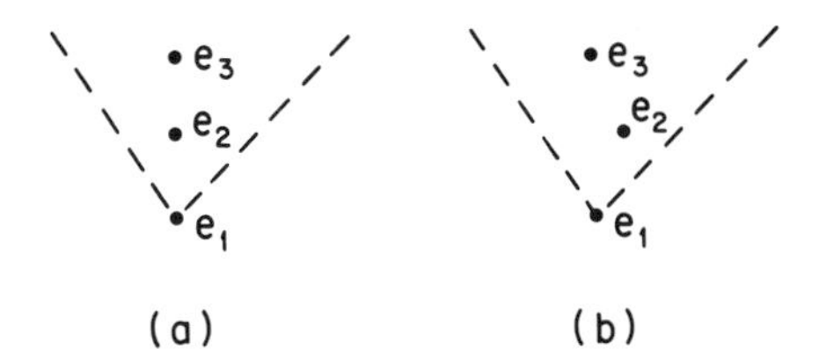

Figure 8.4
Alternative embeddings of events in Minkowski space-time

As with the first uniqueness problem, the relationist can retreat to a conventionalist stance, asserting that whatever is undetermined by the acceptable representations is not a fact. Mundy (1986) rejects conventionalism and commits the relationist to achieving an appropriate expressive equivalence with the absolutist, which he takes to be captured by the condition that whenever ϕ and ϕ' are acceptable embeddings of a relational structure into the same absolutist model, then there is an automorphism h of the model such that $\phi' = h \circ \phi$.[15] This condition is clearly violated in the above example, since the automorphisms of Minkowski space-time preserve relations of collinearity. Thus, as Catton and Solomon (1988) have emphasized, Mundy's example illustrates how the ontological and ideological aspects of the absolute-relational debate are intertwined: ideology depends upon ontology, since whether or not a relation is definable depends upon how rich the domain of the model is.

To achieve expressive equivalence, the relationist may wish to beef up either the ontology or the ideology of his structures. That is, $\mathscr{E}$ could be expanded, as was already contemplated in response to the first uniqueness problem. If there were a physical event at every point of Minkowski space-time, then collinearity among the events would be settled by the causal relations among them. Alternatively, the list of relations holding among the events could be made more inclusive. The crudest expansion would simply add the relation of collinearity to the list—a move some relationists, such as the Reichenbach school, would find uncongenial (see chapter 1).

I find it unfruitful to pursue these issues in terms of such artificial examples. In section 5, I already objected that to make the example more accurate the absolutist models should be expanded to include a set of absolutist events, but even with that emendation the example is wholly unrealistic. What the theories of modern physics give us are not events but field-theoretic models. Of course, absolutist events can easily be constructed from such models, e.g., in electromagnetic-field models we can take a paradigm event to be the event that the Maxwell electromagnetic tensor field takes such and such a value at $p \in M$, but that construction is beside the point. The point is that to be expressively equivalent to realistic absolutist models, the relationist must show how to recover the fields in some interesting fashion. And here a dilemma awaits the relationist. If he sticks to the structureless events $\mathscr{E}$ assumed in the above artificial examples and to spatiotemporal relations among these events, expressive equivalence would seem to be unobtainable in any interesting sense, but the more structure that is added to events, the more they will become like absolutist events, that require the dematerialized ether M for support.

7 What the Antisubstantivalist Must Do

Replacing substantivalist theories of physics with a radically different alternative obviously requires an act of scientific creativity, and no useful advice can be given on how to perform such a feat.[16] But again, I shall register my skepticism about the chances of bringing off the feat if it involves the elimination of the concept of field. Other antisubstantivalists may wish to take seriously what the substantivalist offers and then try to work backward to extract a core content cleansed of substantivalist commitments. For these relationists I can offer some general advice about the contours of the program.

If there were no fields and if all motion were the relative motion of bodies, then it is obvious how to be a consistent instrumentalist about space-time points and yet not succumb to instrumentalism in general: simply embed the relative particle motions into Machian or Leibnizian space-time and regard the embedding as a representation that adds the descriptive fluff of space-time points, as evidenced by the fact that multiple embeddings are available. It is this kind of example that the relationist uses to prime the intuition pump. But the intuition that a nonsubstantivalist theory can be easily constructed is jarred by modern physics. Will the same representationalist ploy allow one to be an instrumentalist with respect to space-time points while being a realist about absolute rotation and acceleration and about fields? In what follows, I will outline the conditions needed to support an interesting positive answer.

The first step is to define an appropriate notion of equivalence for the substantivalist models. We have been assuming that these models have the form $\langle M, O_1, O_2, \ldots \rangle$ where M is the space-time manifold and the O_i are geometric object fields on M. Notice that the O_is may include not only electromagnetic fields and the like but also affine connections and other structures that ground a nonrelational account of motion. Thus, the antisubstantivalist position being explored here is fully prepared to cope with the failure of (R1) and with the adoption of a field ontology. On behalf of the relationist I propose that two such models should be counted as equivalent (*Leibniz-equivalent* I shall say) just in case they can be matched up by a generalized Leibniz-shift operation; that is, $\langle M, O_1, O_2, \ldots \rangle \equiv_L \langle M', O'_1, O'_2, \ldots \rangle$ just in case there is a diffeomorphism d that maps M onto itself so that for all i, $d * O_i = O'_i$. For the antisubstantivalist the intended interpretation of $\mathscr{M} \equiv_L \mathscr{M}'$ is that $\mathscr{M}$ and $\mathscr{M}'$ are different modes of presentation of the same state of affairs; that is, at base, physical states are what underlie a Leibniz-equivalence class of absolutist models.

The second step is to give a direct characterization of the reality underlying a Leibniz-equivalence class. Call this underlying content a *Leibniz model* $\mathscr{L}$. Different relationists will impose on the Leibniz models different tests for relational purity, but presumably all the antisubstantivalists will agree that the $\mathscr{L}$s should not contain space-time points or regions. Reichenbachian relationists will also want to test for ideological as well as ontological purity.

The third step is to demonstrate a kind of expressive equivalence between the Leibniz and the absolutist models. This step can in turn be broken down into two smaller steps. (*a*) There should be a natural sense in which the absolute models are representations of the Leibniz models. The exhibition of the representation relation should be accompanied by a representation theorem showing that under the representation relation, an $\mathscr{L}$ corresponds to a Leibniz-equivalence class of absolute models; that is, if $\mathscr{M}$ represents $\mathscr{L}$, and $\mathscr{M}' \equiv_L \mathscr{M}$, then $\mathscr{M}'$ also represents $\mathscr{L}$, and if both $\mathscr{M}$ and $\mathscr{M}'$ represent the same $\mathscr{L}$, then $\mathscr{M}' \equiv_L \mathscr{M}$. (*b*) The laws of physics should be directly expressible in terms of the $\mathscr{L}$s, and the substantivalist's laws should be recoverable in that if an $\mathscr{L}$ is possible in the lights of the antisubstantivalist laws, then any representation $\mathscr{M}$ of $\mathscr{L}$ is possible in the lights of the substantivalist's laws. This requires that all the members of a Leibniz-equivalence class stand or fall together with regard to nomological possibility—a condition that will be satisfied for generally covariant theories (see chapter 3).

A fourth step is required if the argument of section 4 is accepted. Conditions should be imposed on the Leibniz models that guarantee that for any conforming $\mathscr{L}$, if $\mathscr{M}$ represents $\mathscr{L}$, then the space-time of $\mathscr{M}$ is hole-free.

It is fair to ask why anyone, outside of a few academic philosophers besotted with the absolute-relational controversy, should care about the prospects of this antisubstantivalist program as I have outlined it. In chapter 9 I shall argue that the desire for the possibility of determinism in theories like GTR provides an independent motivation for a program like the above.

It might also be asked why, if steps 1 through 3(*a*) have been carried out, step 3(*b*) cannot be finessed by saying that $\mathscr{L}$ is physically possible just in case it is represented by $\mathscr{M}$s that are physically possible by the substantivalist's lights.[17] This finessing leaves the relationist in the position of claiming that the substantivalist tells a fairy tale using all sorts of fictive entities and at the same time admits that he cannot distinguish between the physically possible and the physically impossible without resort to those fictive entities. This is not a position I would feel comfortable occupying. The debate about the nature of motion is again a useful guiding analogy. If the classical relationists had been forced to admit that empirically adequate laws of motion could not be formulated in terms of relative particle quantities, I am confident that they would not have resorted to finesse but instead would have honestly admitted defeat.[18] I would expect the same honesty on the issue of substantivalism. Of course, it may be that no finesse is needed in that if steps 1 to 3(*a*) are carried out in proper fashion, then step 3(*b*) follows automatically. If, for example, the vocabulary the antisubstantivalist uses to describe the $\mathscr{L}$s is a subvocabulary of the vocabulary used by the absolutist to describe the $\mathscr{M}$s and if the absolutist's theory is axiomatized (the axioms being the absolutist's laws), then the consequences of this theory for the Leibniz states will also be axiomatizable. But the latter axioms may not correspond to anything that is a recognizable scientific law; indeed, in the Craig (1956) reaxiomatization procedure it is clear that most of the axioms will not be lawlike, since there is a separate axiom for each consequence in the privileged subvocabulary. It remains open that some other set of lawlike axioms exists, but that has to be demonstrated. And in any case, it seems unlikely that an interesting representation theorem in steps 1 to 3(*a*) will emerge if the vocabulary of the $\mathscr{L}$s is simply a subvocabulary of the vocabulary of the $\mathscr{M}$s.

If it can be carried out, the upshot of the program outlined above would provide a middle ground between the traditional relational and absolute conceptions of space and time. Perhaps it is this upshot toward which Sklar's maneuver (section 6.9) was striving; if so, Sklar deserves our gratitude for enlarging the possibility space of the discussion. It remains to be seen, however, to what extent such a middle ground is tenable.

8 Conclusion

Those readers who have been keeping score will agree, I trust, that the tally to this juncture is almost wholly in favor of the absolutist position. The relationist loses on the twin issues of the nature of motion and the structure of space-time. And on the issue of substantivalism the relationist is forced into a defensive posture. The absolutist can point to three reasons for accepting a substratum of space-time points: the need to support the structures that define absolute motion, the need to support fields, and the need to ground the right/left distinction when parity conservation fails.[19] Relationists have produced no constructive alternatives to substantivalist theories in physics, and their objections to these theories rest on the theological version of Leibniz's PSR and the verificationist version of the PIdIn, neither of which is compelling. In chapter 6, I examined and rejected Teller's attempt to state a nonverificationist version of Leibniz's argument. Friedman also thinks that Leibniz's objection can be stated in a form that does not amount to verificationism. "The problem," he writes "is not that the absolutist postulates unobservable states of affairs; rather it is that he commits himself to distinct states of affairs that are not distinguishable *even given his own theoretical apparatus*" (1983, p. 219). This is an objection by labeling. For Friedman the states of affairs are "indistinguishable" because they are connected by a transformation that belongs to the "indistinguishability group" of the theory, i.e., the mappings that preserve the space-time structure: the Galilean transformations in the case of neo-Newtonian space-time, the Lorentz transformations in the case of Minkowski space-time. But I do not see why indistinguishability in this sense adds any force to Leibniz's objection, for the appeal to the indistinguishability group is simply a fancy way of restating the construction of Leibniz's Third Letter to Clarke. The "indistinguishable" states of affairs are still separated by properties that any absolutist will want to regard as genuine if he is committed to substantivalism—a commitment he thinks is necessary to adequate accounts of motion, fields, and the left-right distinction. Finally, there remains the possibility of undercutting substantivalism by carrying out the kind of program outlined in section 7. But at present the possibility remains nothing more than that, and so far we have seen no compelling reason to try to turn the possibility into actuality.

Can we then close the book on the absolute-relational controversy? Not quite yet. The most interesting and difficult chapter in the ongoing debate remains to be played out.

9 General Relativity and Substantivalism: A Very Holey Story

Contrary to Reichenbach's claims (chapter 1), relativity theory not only does not vindicate relationism but actually proves to be inimical to relationism in two respects: first, it is hard to square a relational conception of motion with a space-time structure that is recognizably relativistic (see chapter 5), and second, by nurturing the modern field concept, relativity theory also seems to nurture space-time substantivalism (see chapter 8). In the present chapter I shall argue, however, that GTR together with some plausible demands on scientific theorizing deal a blow to one form of substantivalism—what I have called manifold substantivalism—which is at present the only form of substantivalism in the offing. Whether or not some other forms of substantivalism dodge the blow will have to be discussed at length. I begin, in what may initially seem to be a tangential vein, with a bit of sanitized history of GTR.

1 Einstein's Hole Argument

Einstein's GTR was the first successful theory to do without Newton's assumption that the structure of space-time remains "similar and immovable." But the mutability of the structure of space and time is not an idea that is necessarily annexed to general-relativistic space-times or even to relativistic, as opposed to classical, space-times. Contemporary mathematical physicists have constructed all manner of space-times, classical as well as relativistic, in which one bit or another of the structure becomes a dynamic rather than an absolute element. Thus, although what I have to say is, for historical and expository reasons, couched in terms of GTR, neither the physics nor the philosophical moral is unique to GTR.

Having convinced himself that the metric of space-time should be treated as a dynamic element, Einstein began to search for suitable field equations for the "metric field" g_{ik}. His search was guided in part by the *Kausalgesetz* (causal law) that the distribution of matter-energy as specified by the T^{ik} should determine the metric field g_{ik}. Today we would also take it for granted that the field equations should be generally covariant. But Einstein had trouble in finding suitable generally covariant field equations, and by means of an ingenious argument that appeals to the *Kausalgesetz* he managed to convince himself that what he could not do could not be done.[1] Some of the versions of Einstein's argument were fraught with an ambiguity between coordinate and point transformations. The simplified version of

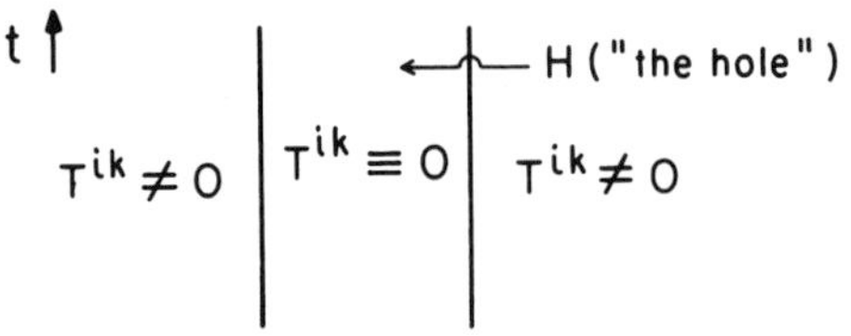

t ↑ H ("the hole")

$T^{ik} \neq 0$ $T^{ik} \equiv 0$ $T^{ik} \neq 0$

Figure 9.1
Einstein's hole construction

the argument that I present here is designed to avoid these and other pitfalls and yet remain true to the direction of Einstein's intentions.[2]

In section 8.4 the "hole" we imagined involved the surgical removal of a portion of the space-time manifold. Einstein's hole argument involves no such metaphysical fantasy but only the existence of a matter-energy hole in the sense of a region H of space-time where the stress-energy tensor vanishes identically (see figure 9.1). If the field equations for g_{ik} and T^{ik} are generally covariant and $\langle M, g, T \rangle$ is a solution, then $\langle M, d * g, d * T \rangle$ is also a solution for any diffeomorphism of M onto itself. We are free to choose d such that $d = \text{id}$ outside the hole H but $\neq$ id inside H and such that the two pieces join smoothly on the boundary of H (a "hole diffeomorphism"). From our assumptions about T^{ik}, it follows that the hole diffeomorphism is a symmetry of the matter-energy distribution; i.e., that $d * T^{ik} = T^{ik}$, which implies that $\langle M, d * g, T \rangle$ is also a solution. We are also free to choose the action of the hole diffeomorphism inside H to be sufficiently non-trivial that $d * g|_H \neq g|_H$. The upshot is that we have produced two solutions, $\langle M, g, T \rangle$ and $\langle M, d * g, T \rangle$, which have identical T fields but different g fields—an apparent violation of the *Kausalgesetz* that the T field determines the g field.

Einstein initially chose to lay the blame on the requirement of general covariance, and for a time he toyed with noncovariant field equations—a move he later acknowledged to be a mistake. But once general covariance is accepted, the other pillar of Einstein's hole argument, the *Kausalgesetz*, comes under fire. And in chapter 6, I argued on independent grounds that the latter requirement is suspect and that it is not implemented in the completed GTR.

It might seem, then, that GTR resolves the problem posed by Einstein's hole argument by incorporating one of the premises of the argument while eschewing the other. If that were all there was to it, Einstein's hole argument would only be a historical quiddity. But even quiddities, if they are Einstein's, can contain the seeds of genius. A slight modification of Einstein's construction shows that no nontrivial form of determinism is possible, at least if manifold substantivalism is taken seriously. In order to appreciate the uniqueness of this threat to determinism, it is necessary first to appreciate, if only to dismiss, more mundane threats.

2 Perils of Determinism

Contrary to the received wisdom, determinism does not find a safe haven in classical physics. To begin with the most obvious problem, it is a commonplace that only in closed systems is it possible to implement the Laplacian brand of determinism according to which the instantaneous state of the system suffices uniquely to fix its future development, for to belabor the obvious, if the system is open to outside influences that have not penetrated the system at the chosen initial moment, no description of the initial state, no matter how detailed, will suffice for a sure prognostication of a future state at a time after which the outside influences penetrate the system. To appreciate the perilous status of classical determinism, it is only necessary to couple the above platitudes with the less obvious remark that even in their entireties, classical universes with space-time structures no stronger than neo-Newtonian space-time (section 2.4) are open systems in the relevant sense. Properly formulated laws of motion in such space-times obey symmetry principle (SP2) (section 3.6), which asserts that the symmetries of space-time are symmetries of the laws. As a result, the laws of motion cannot impose a finite upper bound on the velocity of particle motion, which means that it is possible in principle for a particle such as α in figure 9.2(*a*) to accelerate so strongly that it escapes to spatial infinity without registering on the time slice $t = t^*$. By time-reversal invariance, the temporal mirror image of this process is also physically possible, with the result that a space invader such as β of figure 9.2(*a*) can appear from spatial infinity without leaving its calling card on $t = t^*$. The threat to determinism posed by such processes possible in principle would have only curiosity value unless the processes could be implemented by known forces. A system of point mass particles interacting via Newton's $1/r^2$ law, often cited as *the* paradigm case of classical determinism, may well provide such an implementation.[3]

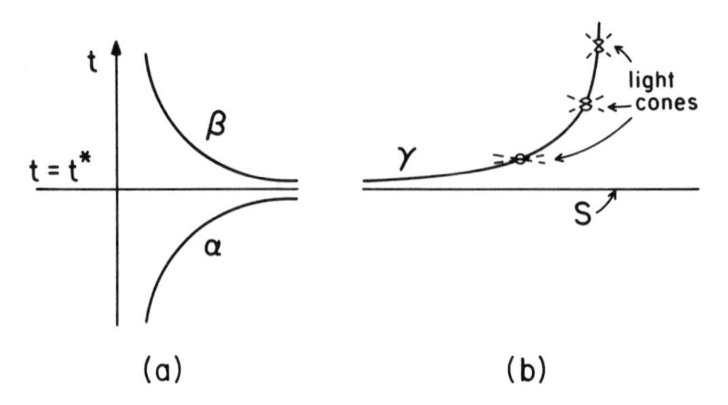

Figure 9.2
Failures of Laplacian determinism

Space invaders are not a problem in STR. The structure of Minkowski space-time, coupled with the postulate that massive particles cannot cross the sub- to superluminal light barrier or vice versa, effectively shields against such processes as α and β of figure 9.2(*a*). However, general relativistic space-time models can exhibit relevant analogues of α and β, as is illustrated in figure 9.2(*b*) which represents the light cone behavior in the universal covering space-time of anti–de Sitter space-time. As one approaches the time slice S from the future, the light cones flatten out, which allows the timelike curve γ to be extended to spatial infinity without ever meeting S.

The presence of D holes (section 8.4) in general-relativistic space-times is another potential pitfall for Laplacian determinism. If one were concerned to give a balanced assessment of the problems and prospects of determinism in GTR, this pitfall and the one sketched above would have to be given serious attention. But since my purpose here is to emphasize the special implications of Einstein's hole argument, I will cavalierly dismiss both pitfalls by fiat by declaring that henceforth I shall restrict attention to general-relativistic models in which the space-time admits a Cauchy surface, a spacelike hypersurface intersected once and only once by every causal curve without endpoint. In the language of section 8.4, that S is a Cauchy surface for M, g means that the domain of dependence $D(S)$ of S is the whole of M.

The final threat to determinism I shall mention here was already covered in chapter 3. From the substantivalist's perspective, the point was that absolute (or immutable) space-time structure has to be sufficiently rich to support any interesting form of determinism. Specifically, it seems to follow from the application of (SP2) that neither Machian, Leibnizian, nor Maxwellian space-time can support laws of particle motion that will determine future motions from past motions. Two types of responses were contemplated. Those inclined against relationism will be happy to shore up the prospects of Laplacian determinism by beefing up the structure of space-time by, say, moving to neo-Newtonian space-time, for such a move also undermines the relational account of motion. Those inclined toward relationism will want to stick with the more lightly structured space-times and abandon the substantivalist interpretation of symmetry principles (underlying the argument from [SP2]) as moving particle systems around in the space-time container. In a context where the space-time structure remains "similar and immovable," I do not see any decisive consideration for or against one of these alternative reactions. But when the space-time structure becomes mutable, the substantivalist tactic of adding more structure is of no help in salvaging determinism, or so I will now try to show.

3 Space-Time Substantivalism and the Possibility of Determinism

Choose any general relativistic model $\langle M, g, T \rangle$ you like, except that, in keeping with the sentiments of section 2 for making the environment as friendly as possible for determinism, suppose that M, g possess a Cauchy surface. This supposition implies that there is a global time function $t: M \rightarrow \mathbb{R}$ such that t increases as one moves in the future direction along any timelike curve and such that the level surfaces of t are all Cauchy surfaces. Following the style of Einstein's hole construction, choose a diffeomorphism d such that $d = \text{id}$ for all $t \leq 0$ and $\neq \text{id}$ for $t > 0$ and such that there is a smooth join at $t = 0$. By general covariance, $\langle M, d * g, d * T \rangle$ is also a model, and M, $d * g$ also possesses a Cauchy surface. By construction, this new model is identical with the first for all $t < 0: g|_{t \leq 0} = d * g|_{t \leq 0}$ and $T|_{t \leq 0} = d * T|_{t \leq 0}$. But if d is chosen properly, the models will differ for $t > 0$—a seeming violation of the weakest form of Laplacian determinism.[4]

Nor is Laplacian determinism the only form of determinism that suffers; indeed, any nontrivial form of determinism suffers equally. For example, choose any closed $N \subset M$, no matter how small. By an obvious modification of the above construction, it would seem to follow also that the state throughout the rest of space-time $M - N$ cannot fix the state in N.

The incompatibility demonstrated here between the possibility of determinism and space-time substantivalism holds for any form of substantivalism that implies that $\langle M, g, T\rangle$ and $\langle M, d*g, d*T\rangle$ are distinct when $d*g \neq g$ and $d*T \neq T$. This is certainly the case for what I call manifold substantivalism, which is the only form of substantivalism presently in the offing. That view can be characterized in terms of political economics as an exercise in the division of labor. The differential geometer constructs the Ms and then passes them on to the physicist who proceeds to test them for suitability as a basis for a general-relativistic model: Is M paracompact? Can M carry a nonsingular Lorentz metric? etc. Of those that are suitable, some will go on to be used as arenas for solutions to Einstein's field equations of gravitation. But in any case it is assumed that questions of identity and individuation of points of M have been settled prior to the introduction of the g-field and the T-field; indeed, the very characterization of fields given in section 8.3 takes for granted the identity of the elements in the point set, the topology on the set, and the differential structure. Then, since the fields introduced on M have physical significance, shifting those fields will produce different physical states of affairs just as shifting particles in three-space produces different physical states for the space substantivalist. This is not to say that there cannot be some other form of substantivalism that is reconcilable with determinism within theories like GTR, but I shall postpone the question until sections 11 to 14.

Before closing this section, I should note that what is true of Leibniz's original argument is also true of the present argument: if the argument works against an ontology of irreducible and ineliminable space-time points, it also works against an ontology of irreducible and ineliminable monadic properties of spatio-temporal locations.[5] The difference between Leibniz's argument in the correspondence with Clarke (see chapter 6) and the present one is that considerations of determinism have more polemical force than Leibniz's considerations of verificationism and the nomenclature of God's decision-making process. I shall defend this judgment of differential merits in the following section.

4 Taking the Possibility of Determinism Seriously

The argument in the preceding section does not rest on the assumption that determinism is true, much less on the assumption that it is true *a priori*, but only on the presumption that it be given a fighting chance. To put the matter slightly differently, the demand is that if determinism fails, it should fail for a reason of physics. We have already seen a couple of such reasons. For example, the laws of physics might allow arbitrarily fast causal signals, as in Newtonian mechanics, opening the way for space invaders, or the laws might impose a finite upper bound on causal propagation but, as in some general-relativistic cosmological models, still permit analogues of space invaders. Alternatively, the laws might forbid space invaders but, as in some interpretations of quantum mechanics, admit irreducibly stochastic elements. Or less interestingly, the actual laws might be weaker than anything we currently take to be laws, so weak that the present state fails to place effective lawlike constraints on future states. But the failure of determinism contemplated in section 3 is of an entirely different order. Rule out space invaders; make the structure of space-time as congenial to determinism as you like by requiring, say, the existence of a Cauchy surface; ban stochastic elements; pile law upon law upon law, making the sum total as strong as you like, as long as logical consistency and general covariance are maintained. And still determinism fails if the proffered form of space-time substantivalism holds. Here my sentiments turn against substantivalism.

Having appealed to sentiments, I now stoop to appeal to authority. Relativity physicists are, of course, aware of the construction of section 3. But to my knowledge not one of them is on record as concluding that determinism fails in GTR. It has to be admitted, however, that the upshot of the construction is left in doubt in the older-style textbooks because of the ambiguity between coordinate and point transformations.

Thus, consider the way the construction emerges in the midst of Møller's (1952, pp. 310–312; 1972, pp. 424–425) discussion of properties of the gravitational field equations. His starting point is that the equations are of the form

$$M_{ik} \equiv R_{ik} + c_1 R g_{ik} + c_2 g_{ik} = -\kappa T_{ik}, \tag{9.1}$$

where R_{ik} is the Ricci tensor and R is the scalar curvature. Taking into account the four-dimensionality of space-time and the fact that $M_{[ik]} = T_{[ik]} = 0$, we can see that (9.1) gives ten second-order partial differential equations for the g_{ik}. But only six of these equations can be independent. To see why, take the empty-space case where $T_{ik} \equiv 0$ (the "big hole"). If the resulting field equations

$$M_{ik} = 0 \tag{9.2}$$

were independent, then in a coordinate system x^i the functions $g_{ik}(x^l)$ would be uniquely determined by the values of the g_{ik} and their first derivatives on the initial-value hypersurface $x^4 = 0$. But, Møller continues, such a unique determinism is impossible. Assume for purposes of contradiction that it is possible, and introduce a new coordinate system where the coordinate transformation is indicated by

$$x'^j = x'^j(x^i), \qquad x^i = x^i(x'^j). \tag{9.3}$$

By general covariance, M'_{ik} is the same function of g'_{ik}, $\partial g'_{ik}/\partial x'^l$, and $\partial^2 g'_{ik}/\partial x'^l \partial x'^m$ as M_{ik} is of g_{ik}, $\partial g_{ik}/\partial x^l$, and $\partial^2 g_{ik}/\partial x^l \partial x^m$. And again by general covariance, if (9.2) is true, then so is

$$M'_{ik} = 0. \tag{9.4}$$

So if they were independent, the equations (9.4) would uniquely determine the functions $g'_{ik}(x'^l)$ from the values of g'_{ik} and their first derivatives on $x^4 = 0$. Now choose the new coordinates so that $x'^i = x^i$ on and near the initial value hypersurface but $x'^i \neq x^i$ elsewhere. By the first part of the choice, $g'_{ik} = g_{ik}$ and $\partial g'_{ik}/\partial x'^l = \partial g_{ik}/\partial x^l$ on the initial value hypersurface, with the result that the g'_{ik} must be the same functions of x'^l as the g_{ik} are of x^l. But this contradicts the second part of the choice, since

$$g'_{ik} = \frac{\partial x^l}{\partial x'^i} \frac{\partial x^m}{\partial x'^k} g_{lm}. \tag{9.5}$$

Møller concludes that the ten equations (9.2) cannot be independent but must satisfy four identities. “This means that the solutions g_{ik} of the field equations (9.2) contain four arbitrary functions corresponding to the four arbitrary functions in the transformations (9.3), which only change our space-time description, but not the physical system which produces the gravitational field” (1972, p. 425).[6]

The lineage of Møller’s argument traces back to Einstein’s hole argument. Møller credits Hilbert,[7] and Hilbert in turn had read Einstein’s papers and most likely had heard the hole argument first hand from a lecture Einstein gave in Göttingen in 1915.[8] But Møller’s version leaves the upshot in doubt. The field equations do not uniquely determine the g_{ik} as functions of x^l, but that underdetermination, Møller seems to indicate, is not a failure of determinism, since the coordinate transformations “only change our space-time description, but not the physical system.” But is that because (trivially) different coordinate components of the same g-field in different coordinate systems describe the same physical situation, or because (more interestingly) different but diffeomorphically related g-fields describe the same physical situation? Other textbooks likewise leave the question hanging. After going through the same construction, Adler, Bazin, and Schiffer (1975) conclude that

> The initial data on S [the initial-value hypersurface] do not determine the resulting metric in a unique way; the solution contains four arbitrary functions $\partial^2 g_{i4}/\partial x^4 \partial x^4$ which are at our disposal. It should be observed that this arbitrariness is due to the fact that we can pick an arbitrary coordinate system for the description of the space-time continuum. However, the solutions obtained will differ only formally; they will describe the same geometrico-physical situation in different reference systems.[9] (1975, p. 279)

More recent textbooks leave no room for equivocation. Hawking and Ellis 1973, for example, states that in the initial-value problem in GTR, one can expect uniqueness only "up to a diffeomorphism."

5 The Role of the Mutability of Space-Time Structure

I have given the impression that the mutability of space-time structure entailed by GTR plays a key role in generating the conflict between substantivalism and the possibility of determinism. One might question that impression by noting that special relativistic theories can be written in forms that apparently allow the construction of section 3 to be applied. For example, take a special relativistic theory of motion and rewrite the equations using covariant derivatives with respect to an undetermined Lorentz metric g. Then write a "field equation" for g, namely,

$$R^i{}_{jkl}(g) = 0, \tag{9.6}$$

where $R^i{}_{jkl}$ is the Riemann curvature tensor. This maneuver appears to let one have one's cake and eat a large slice of it too. Equation (9.6) and the requirement that the manifold be $\mathbb{R}^4$ entail that the space-time is Minkowskian, as required by the standard interpretation of STR. But it also appears to allow one to say that the space-time metric is not given *ab initio* but is determined, to the extent that it is determined at all, in the same way that other physical variables are determined. A similar maneuver can also be applied to classical space-time theories.

While such formalism is unassailable *qua* formalism, its uncritical use papers over some important distinctions. With respect to classical and special-relativistic space-times it is possible to be a hypersubstantivalist and maintain that there is just one space-time and that talk about different worlds is to be translated as talk about different arrangements of matter and fields in the fixed space-time. I take it that this was Newton's attitude in "De gravitatione," and the attitude is consistent with the passage from Truesdell quoted in the appendix to chapter 2. The point is that the hypersubstantivalist who takes the structure of the fixed space-time to be strong enough—e.g., neo-Newtonian or Minkowskian—can have both substantivalism and determinism. The problem is that such hypersubstantivalism seems to go against the requirement of general covariance as used above and in previous chapters.

Formalism generated the problem, and formalism is needed to resolve it. I shall suppose that we have made a distinction between absolute and dynamic objects and that this distinction corresponds to the distinction between the object fields (A_i) that characterize the structure of space-time and those (P_j) that characterize the physical contents of space-time. The space-time structure need only be assumed to be absolute in the following minimal sense: for any two dynamically possible models of the theory $\mathcal{M} = \langle M, A_1, A_2, \ldots, P_1, P_2, \ldots \rangle$ and $\mathcal{M}' = \langle M', A'_1, A'_2, \ldots, P'_1, P'_2, \ldots \rangle$, there is a diffeomorphism $d: M \rightarrow M'$ such that $d * A_i = A'_i$ for all i. (This is the "remains similar" requirement of the appendix to chapter 2.) I shall call such a d an absolute map. We can then say that the theory is minimally Laplacian-deterministic just in case for any models $\mathcal{M}$ and $\mathcal{M}'$ and any absolute map d, if $d * P_j|_{d*t \leq 0} = P'_j|_{d*t \leq 0}$ for all j, then $d * P_j = P'_j$ for all j, where $t = 0$ is a plane of absolute simultaneity or a Cauchy surface of the space-time of $\mathcal{M}$.

Generally covariant special-relativistic theories using Minkowski structure or generally covariant classical theories using neo-Newtonian space-time structure can be minimally deterministic, but the substitution of weaker structures such as those of Machian, Leibnizian, or Maxwellian space-time (sections 2.1 to 2.3) undermines determinism. In addition, moving an item from the A side to the P side of the absolute-dynamic cut can also change the fortunes of determinism. Thus, in Cartan's formulation of Newtonian gravitation the affine connection ${}^{C}\Gamma$ is in general not flat, and more to the point, it varies from model to model, with its values being determined by a field equation that relates ${}^{C}\Gamma$ to the mass distribution. In the language of the appendix of chapter 2, ${}^{C}\Gamma$ neither remains similar nor remains immovable, and so we are justified in putting it on the P side of the cut. The result is that an absolute map now has only to match up the planes of absolute simultaneity, the $\mathbb{E}^3$ structure of the planes, and the intervals between nonsimultaneous events. Such a map still contains enough freedom to violate minimal Laplacian determinism, at least if space-time substantivalism is adhered to. The extreme case occurs when all the As are erased, with only Ps remaining. In that case substantivalism leads to the demise of any interesting form of determinism. That is the case of GTR.

6 The Leibnizian Reaction

I am morally certain that Leibniz would have endorsed the argument of sections 3 and 4; indeed, I think that he would have claimed it as his own.[10] The claim has some plausibility. Although Leibniz never advocated the mutability of space-time structure, mutability is a comfortable companion to his doctrine that space is the order of relations of coexistences and time is the order of relations of successive events. And more important, the core of the argument in sections 3 and 4 consists of an application of the causal version of PSR. In his Second Reply, Clarke conceded, " 'Tis very true, that nothing is, without a sufficient reason why it is, and why it is thus rather than otherwise. And therefore, where there is no cause, there can be no effect" (Alexander 1984, p. 20). Applying the "where there is no cause, there can be no effect" version of PSR to the case of GTR, Leibniz could argue that the difference in the g- and T-fields after $t = 0$ must be traceable to differences before $t = 0$. But by construction, there is no difference prior to $t = 0$. Thus, to avoid a violation of the causal version of PSR, substantivalism must be abandoned. I would only add that the specific form of the argument given in sections 3 and 4 is stronger than the one I have put in Leibniz's mouth, for my version allows that the differences after $t = 0$ might be traceable not to a difference before $t = 0$ but to space invaders, stochastic elements, etc., but it shows that even if these and other defeasors are ruled out, substantivalism is inconsistent with "where there is no cause, there can be no effect."

Finally, the obvious way to save the causal version of PSR is to say that the two substantival models constructed in section 3 are just different modes of presentation of the same physical state of affairs, and it is only a short step from there to say that any two diffeomorphically related substantival models provide equivalent descriptions. Since this notion is clearly a straightforward extension of Leibniz's treatment of the different substantivalist models obtained by shifting bodies in space (see sections 6.2 to 6.5), I have called the equivalence Leibniz equivalence in anticipation (see section 8.7).

Modern general-relativists come close to adopting a Leibnizian stance, although there is admittedly room for equivocation. Thus, Hawking and Ellis write, "Strictly speaking then, the model for space-time is not just one pair (M, g) but a whole equivalence class of pairs (M', g') which are equivalent to (M, g)" (1973, p. 56). (Their notion of equivalence of space-time models is just the one I have been using above.) And in a similar vein, Wald writes:

> If $\phi : M \rightarrow N$ is a diffeomorphism, then M and N have identical manifold structure. If a theory describes nature in terms of a space-time manifold, M, and tensor fields, $T(i)$, defined on the manifold, then if $\phi : M \rightarrow N$ is a diffeomorphism, the solutions $(M, T^{(i)}$ and $(N, \phi * T^{(i)})$ have physically identical properties. Any physically meaningful statement about $(M, T^{(i)})$ will hold with equal validity for $(N, \phi * T^{(i)})$. (Wald 1984, p. 438)

Of course, to say that two models are "physically equivalent" or "have physically identical properties" is ambiguous between (1) corresponding to distinct but physically indistinguishable states of affairs, and (2) giving different descriptions of the same state of affairs. But I take it that the endorsement of determinism for Einstein's field equations favors (2).

The next obvious question is what is entailed by taking (2) seriously. But before turning to that matter, it is worth pausing to survey Einstein's own considered reaction to his hole argument.

7 Einstein's Reaction

Since Einstein's hole argument set off the line of inquiry pursued in this chapter, it is only natural to wonder how he came to assess the implications of his argument. In particular, after he reaffirmed general covariance and after he had realized that the dialectic of the argument leads to a widespread underdetermination, not just to an underdetermination of the g-field by the T-field, how did he read the significance of the underdetermination?

The first answer is more than a little disappointing in its reliance on a crude verificationism and an impoverished conception of physical reality. In the paper that laid the foundations of the final GTR, Einstein wrote:

All our space-time verifications invariably amount to a determination of space-time coincidences. If, for example, events consisted merely in the motion of material points, then ultimately nothing would be observable but the meeting of two or more of these points. Moreover, the results of our measurings are nothing but verifications of such meeting of the material points of our measuring instruments with other material points, coincidences between the hands of a clock and the points on a clock dial, the observed point-events happening at the same place at the same time. (1916, p. 117)

Now since all of our physical experiences can be reduced to such coincidences and since all physical laws are just summaries of patterns of such experiences, it follows that the freedom to perform diffeomorphisms does not lead to an underdetermination in physics, for diffeomorphisms preserve such coincidences. This is not quite the language Einstein used in his 1916 paper—he was using the terminology of coordinate systems and coordinate transformations—but the basic thrust emerges nonetheless:

The introduction of a system of reference serves no other purpose than to facilitate the totality of such coincidences. We allot to the universe four space-time variables x_1, x_2, x_3, x_4 in such a way that for every point-event there is a corresponding system of values of the variables.... To two coincident point-events there corresponds one system of values of the variables $x_1 \ldots x_4$, i.e. coincidence is characterized by the identity of the coordinates. If, in place of the variables $x_1 \ldots x_4$, we introduce functions of them x'_1, x'_2, x'_3, x'_4, as a new system of coordinates, so that the systems of values are made to correspond to one another without ambiguity, the equality of all four coordinates in the new system will also serve as an expression of the space-time coincidence of the two point-events. (1916, pp. 117–118)

Lest the reader think that these remarks of Einstein's are not to be taken seriously because they are only introductory filler to the serious part of his paper, it is worth quoting from a letter Einstein wrote in January of 1916 to his friend Paul Ehrenfest. Having previously convinced Ehrenfest of the need to abandon general covariance, Einstein was then in the curious position of having to unconvince him. Einstein's persuasion was couched in terms of an example of Ehrenfest's involving the illumination of a photographic plate by means of starlight that passes through an aperture.

Your example somewhat simplified: you consider two solutions with the same boundary conditions at infinity, in which the coordinates of the star, the material points of the aperture and of the plate are the same. [See figure 9.3(*a*).] You ask whether "the direction of the wave normal" at the aperture always comes out the same. As soon as you speak of "the direction of the wave normal *at the aperture*," you treat this space with respect to the functions $g_{\mu\nu}$ as an infinitely small space.

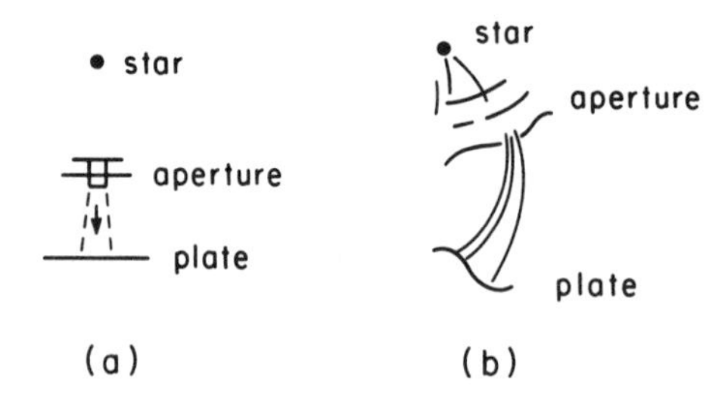

Figure 9.3
Einstein's thought experiment

From this and the determinateness of the coordinates of the aperture it follows that the direction of the wave normal AT THE APERTURE *for all solutions are the same.*

This is my thesis. For a more detailed explanation [I offer] the following. In the following way you recover all solutions allowed by general covariance in the above special case. Trace the above little figure onto completely deformable tracing paper. Then deform the paper arbitrarily in the plane of the paper. Then make a carbon copy back onto the writing paper. Then you recover e.g. the figure [fig. 9.3(*b*)]. When you relate the figure once again to orthogonal writing paper coordinates, the solution is mathematically different from the original, and naturally also with respect to the $g_{\mu\nu}$. But physically it is exactly the same, since the writing paper coordinate system is only something imaginary. The same point of the plate always receives the light....

The essential thing is: as long as the drawing paper, i.e. "space," has no reality, then there is no difference whatever between the two figures. It [all] depends on coincidences.[11]

The second answer is more interesting but also more cryptic. In the fifteenth edition of *Relativity: The Special and the General Theory*, Einstein added an appendix that showed he had dropped the narrowly positivistic conception of physical reality assumed in his first response, in favor of a view that accords fields, especially the g-field, a basic role:

If we imagine the gravitational field, i.e., the functions g_{ik}, to be removed, there does not remain a space of the type (1) [Minkowski space-time], but absolutely *nothing*, and also no "topological space." For the functions g_{ik} describe not only the field, but at the same time also the topological and metrical structural properties of the manifold.... There is no such thing as an empty space, i.e., a space without field. Space-time does not claim existence on its own, but only as a structural quality of the field. (1961, p. 155)

Philosophers of every persuasion have claimed to nestle under Einstein's wing, and it is no different in the absolute-relational debate. Relationists will no doubt see Einstein's dictum that "space-time does not claim an existence of its own, but only as a structural quality of the field" to be an endorsement of antisubstantivalism. How such an antisubstantivalism is to be realized, Einstein did not say, but his remarks could be taken to suggest the view to be explored in section 9 below. On the other hand, substantivalists might also take comfort from Einstein's remarks, as we shall see in sections 11 and 12. But before turning to details, it will be useful to structure the discussion by setting out a taxonomy of possible reactions to Einstein's hole construction.

8 A Catalog of Responses to Einstein's Hole Construction

In what follows I shall ignore instrumentalist responses, such as Einstein's initial idea that no real form of underdetermination is involved, because all of physical reality reduces to space-time coincidences such as the intersection of light rays. The common thread to all of the noninstrumentalist reactions discussed below is a realism about fields in general and about the metric field in particular. But this thread is much too slender to stitch together any significant unity, for the noninstrumentalist responses are sharply divided between the relationist and absolutist camps, and within the latter there is an untidy variety of subcamps.

The relationist will wish to urge that the hole construction, which may be viewed as a causal version of Leibniz's PSR argument, is polemically effective where Leibniz's original theological version was not. That is, the construction is to be interpreted as showing that if one is a realist about fields in the context of theories like GTR, then one must be an instrumentalist about space-time points. The main challenge to this point of view is the apparent need to use space-time as support for fields. Two possible responses to this challenge are discussed below in sections 9 and 10.

In reaction to the hole construction the absolutist may contemplate a partial retreat or may opt for an uncompromising hold-the-line response. The strategy of partial retreat would involve giving up space-time points but would maintain that space-time remains substantival at some deeper level. Reasons for this sentiment will emerge in sections 9 and 10. The no-retreat strategy would continue to maintain a realism with respect to space-time points either by defending manifold substantivalism or by replacing it with some other form of space-time substantivalism.

The first fork of the no-retreat strategy can be implemented by a tactic that seeks to absorb the blow of the hole construction or alternatively by a tactic that seeks to dodge the blow. The absorption tactic would cheerfully admit that the hole construction shows that determinism fails but would claim that this failure is far from fatal to manifold substantivalism.[12] The failure is certainly not as pernicious as, say, that involved in an irreducibly stochastic process. Moreover, general-relativists can be taken as endorsing this point of view when they say that the initial-value problem in GTR can have a unique solution only "up to a diffeomorphism." However, as seen in section 6 above, the very same physicists want to overcome the underdetermination problem by taking Leibniz-equivalence classes of absolutist models and claiming that they correspond one to one to physical states. This tension suggests that the tactic of trying to absorb the blow is not a happy one.

Dodging the blow would be achieved by showing that, contrary to first appearances, no underdetermination is involved in the hole construction. Taking the cue from the discussions of Møller and of Adler, Bazin, and Schiffer reviewed above in section 4, one could hold that all gs produced in the hole construction differ only formally and describe the same geometric physical situation. This tactic would be understandable on a passive construal of the hole construction, for then $d_1 * g$, $d_2 * g, \ldots$ for various ds would stand not for different metrics but merely for different presentations of the same metric in different coordinate systems, and thus there would be no more underdetermination in physics than there is in the expression of the mass of a particle in different systems of units. But on the active interpretation of the construction, there are in the offing many different metrics that, according to manifold substantivalism, predict objectively different properties of space-time points; for example, $d_1 * g$ may say that points p and q are relatively lightlike, while $d_1 * g$ says that they are relatively spacelike, which leads to the conflicting predictions that p and q can and cannot be connected by a nonbroken light ray.

The appearance of attributing contradictory properties can perhaps be avoided by taking space-time points to possess multiple conjunctive properties each of which is multiply conditional;[13] namely, point p carries $g(p)$ when q carries $g(q)$ and r carries $g(r)$, etc. through all the points; p carries $d_1 * g(p)$ when q carries $d_1 * g(q)$, etc.; and so on through all the diffeomorphisms. One can wonder, however, how the substantivalist can use such properties to explain the phenomena without also invoking some determinate metrical properties of space-time points and thereby unraveling the web of conjunctive conditionals. One can also wonder whether such conditional properties ought to be countenanced in the discussion of determinism. For if they are admitted, it may be hard to find a principled way to exclude such properties as 'futurizes up (down)' which now applies to a spin $\frac{1}{2}$ particle just in case it emerges spin up (down) from a Stern-Gerlach apparatus the first time in the future it is run through. With the admission of such properties, determinism seems to dissolve into a triviality.

If manifold substantivalism cannot be sustained under the pressure of the hole construction, there are still three ways to uphold substantivalism with respect to space-time points: adopt a structural-role theory of identity for space-time points, claim that metrical properties are essential to space-time points, or apply counterpart theory to space-time models. These three tactics will be discussed in turn in sections 12, 13, and 14.

9 Trying to Do without Space-Time Points

One way to ground the idea that substantivalist models are merely different representations of the same physical reality is to carry out the program outlined in section 8.7. In view of the discussion in this chapter, one should add to the list of requirements on the Leibniz models the further demand that they make determinism possible. I now propose to argue that although the relationist can carry out the program part way by eliminating space-time points, it is dubious that the program can be pushed so far as to eliminate all substantivalist entities, at least as long as the field concept is retained in some recognizable form.

As an example of how points can be eliminated at one level but recovered in terms of representations at another level, let us start with a topological space X, consisting of a point set and a collection of open sets. Consider the collection $C_0(X)$ of all continuous real-valued functions on X. Equipped with the operations of pointwise addition and multiplication, $C_0(X)$ forms a ring. The subring of all bounded continuous functions is denoted by $C_0^*(X)$, and if X is compact, $C_0^*(X) = C_0(X)$. Once we have constructed the rings, we can throw away the underlying point space and consider C_0 and C_0^* as algebraic objects in their own right. These algebraic objects can in turn be represented and realized respectively as rings of continuous and bounded continuous functions on a topological space. Many such realizations of the same algebraic object are possible, but if the representing spaces have some nice properties, the algebraic object determines the representing space up to topological equivalence. For if X and X' are two bases of realization of the same algebraic object, then $C_0(X)$ and $C_0(X')$ are isomorphic as rings, and likewise for $C_0^*(X)$ and $C_0^*(X')$. And if X and X' are compact, the isomorphism of $C_0(X)$ and $C_0(X')$ implies that $X \cong X'$ (i.e., X and X' are homomorphic). And if X and X' are completely regular and satisfy the first axiom of countability, conditions satisfied by spaces used in physics, then the isomorphism of $C_0^*(X)$ and $C_0^*(X')$ again implies that $X \cong X'$ (see Gillman and Jerison 1960 and Nagata 1974).[14] In sum, for a broad category of topological spaces there is an algebraic structure common to topologically equivalent spaces, and this structure is in turn strong enough to determine the topology. Thus, one can view the basic objects of analysis as algebraic objects and treat different but equivalent topological spaces as merely different representations of the same basic object.

Geroch (1972) showed how to extend these ideas to general relativity and similar theories. A C^∞ manifold M allows us to define in addition to $C_0(M)$ and $C_0^*(M)$ two other rings, $C_\infty(M)$ (the ring of C^∞ real-valued functions), and $C_c(M)$ (the ring of constant functions), which is isomorphic to $\mathbb{R}$. As we saw in section 8.3, a smooth contravariant vector field on V can be characterized as a mapping from $C_\infty(M)$ to $C_c(M)$ such that $V(\lambda f + \mu g) = \lambda V(f) + \mu V(g)$ and $V(fg) = fV(g) + V(f)g$ for $f, g \in C_\infty(M)$ and $\lambda, \mu \in C_c(M)$. Covariant tensor fields can then be characterized as multilinear maps from tuples of contravariant vector fields to $C_c(M)$. And in fact, all the machinery needed to do general-relativistic gravitation can be coded up in terms of such operations. Now in parallel with the move made for topological spaces, throw away the manifold M and keep the algebraic structure, which we may call a *Leibniz algebra*.[15] Such an $\mathscr{L}$ can be realized or represented by a standard model $\langle M, g, T \rangle$ in many ways, but any two such realizations of the same $\mathscr{L}$ will be Leibniz-equivalent.[16] Thus, it is open to take the $\mathscr{L}$s as giving a direct characterization of physical reality and view the different members of the corresponding Leibniz-equivalence class of substantival models as giving different representations of the same reality. This provides one plausible reading, I think, of Einstein's idea that space-time, M, "does not claim an existence of its own, but only as a structural quality of the field."

While the Leibniz algebras provide a solution to the problem of characterizing the structure common to a Leibniz-equivalence class of substantival models and the solution eschews substantivalism in the form of space-time points, the solution is nevertheless substantival, only at a deeper level. And the Leibniz algebras do not escape the theological version of Leibniz's PSR objection, for while it is true that all the members of a Leibniz-equivalence class of substantival models can be recovered as realizations of a single $\mathscr{L}$, it is equally true that the different members of the equivalence class can be taken to generate different but isomorphic Leibniz algebras. The proliferation of equivalent substantival models is thus matched by an equinumerous proliferation of equivalent Leibniz algebras, and the problem of why God should choose to actualize one rather than another of the former is transferred to the problem of why He should choose to actualize one rather than another of the latter.

As noted in chapter 6, no theory formulated in a standard logical language can hope to fix the models more uniquely than up to isomorphism. So the existence of isomorphic models does not by itself impugn the theory or give grounds for doubting the existence of the entities postulated by theory. To generate such qualms, an additional consideration must be supplied. In the case of theories presupposing manifold substantivalism, the relevant consideration is the demand for the possibility of determinism. In the case of the Leibniz algebras the desire to secure the possibility of determinism does not add any additional pressure to find a purely relational solution to the PSR problem, for the proliferation of Leibniz algebras does not threaten determinism. In keeping with the spirit of the present approach, the laws of nature must be expressed directly in terms of the Leibniz algebras, but the meaning of determinism is to be understood in terms of the space-time realizations of the algebras. Suppose that the laws of nature have spoken and that the physically possible Leibniz algebras have been picked out. Then in its minimal form, Laplacian determinism demands that for any physically possible $\mathscr{L}$ and $\mathscr{L}'$, the class of space-time representations of $\mathscr{L}$ is identical to the class of space-time representations of $\mathscr{L}'$ if the restrictions to $t \leq 0$ of the classes of representations are identical.[17] Whether or not this property holds is a contingent matter, but nothing about the construction of the $\mathscr{L}$s precludes the possibility that it holds.

For those who want a more thoroughgoing relationism, it would be necessary to exorcise from the Leibniz algebras the ghost of substantival space-time—the prespace, as it were—or else to follow some rather different path. The former route seems to me to be unpromising, at least if the concept of field is retained and interesting representation theorems are to remain provable. Another path to relationism is explored in the next section.

10 Another Relationist Approach

Another more thoroughly relational and more constructivist approach would proceed in two steps: first, the space-time manifold would be built up from physical events and their relations, and then with the manifold in hand the characterization of fields could proceed as usual. To implement the crucial first step, we start with a plenum of physical events $\mathscr{E}$. (From the substantivalist's point of view, $\mathscr{E}$ would consist of punctal events that cover M; i.e., for any $p \in M$, there is an $e \in \mathscr{E}$ such that e occurs at p. Of course, the relationist cannot in good conscience take this as a definition of plenitude.) And we suppose that $\mathscr{E}$ is equipped with a set of relations, which for definiteness we, following the Reichenbach school, take to be causal relations. In particular, consider the binary relations t, c, l of timelike, causal, and lightlike precedence. (From the substantivalist's perspective, $t(e_1, e_2)$ [respectively, $c(e_1, e_2)$, $l(e_1, e_2)$] just in case e_1 and e_2 occur respectively at points $p_1, p_2 \in M$ and there is a future-directed timelike (respectively, causal, lightlike) curve from p_1 to p_2.[18] But for the relationist these are primitive relations, and the substantivalist's "explanation" of these relations is turned on its head.) Two events $e_1, e_2 \in \mathscr{E}$ are said to be causally equivalent just in case for any $e_3 \in \mathscr{E}$, $t(e_1, e_3)$ if and only if $t(e_2, e_3)$, $t(e_3, e_1)$ if and only if $t(e_3, e_2)$, and similarly for c and l. A space-time point is then defined as a causal-equivalence class of events, and the causal relations t, c, and l are transferred in the obvious way to relations $\bar{t}, \bar{c}, \bar{l}$ on the set of points P(e.g., $\bar{t}(p_1, p_2)$ if and only if there are e_1 and e_2 in the equivalence classes of events corresponding respectively to p_1 and p_2 such that $t(e_1, e_2)$). A topology is then induced on P by taking as a basis for open sets, sets of the form $I^+(p) \cap I^-(q)$, p, $q \in P$, where $I^+(p) \equiv \{p' \in P : \bar{t}(p, p')\}$ and $I^-(q) \equiv \{q' \in P : \bar{t}(q', q)\}$.

From the substantivalist's perspective the construction of points will be satisfactory only if distinct space-time points have distinct causal pasts and/or futures. And the definition of topology will be satisfactory only if the space-time is strongly causal (i.e., for every $p \in M$ and every neighborhood of p, there is a subneighborhood that is not intersected more than once by any timelike (curve), for then and only then will the causal topology coincide with the manifold topology (see Hawking and Ellis 1973, pp. 196–197). Since there are various general-relativistic cosmological models that violate these causality conditions, the relationist will have to advance considerations that show such models are beyond the pale of physical possibility. It is sometimes claimed that the causal paradoxes that would result from closed timelike curves demonstrate the conceptual absurdity of such things. I think that this is a mistake,[19] and in any case a violation of strong causality can occur without there being closed causal curves.

Furthermore, even accepting the above causal construction of topology, I do not see how to continue the construction to include the differential structure. And the entire procedure gives off whiffs of circularity. It is one thing to start with events of the type Einstein imagined (e.g., coincidences such as the collision of two particles), but it is quite another to resort, as presumably one must to get a plenum of events, to events that the absolutist takes to be the tokening of values of fields at space-time points and the relationist takes as unanalyzed primitives.

I do not want to be taken as suggesting that the type of program alluded to above is not worth pursuing; on the contrary, I think we have much to learn from it whether or not it succeeds. But I would again note the recurring point that relational theorists leave more promissory notes than completed theories.

11 An Absolutist Counterattack

The absolutist might rest content with trying to establish what I left as conjectures in sections 9 and 10, namely, that the models of reality produced in the program of section 9 remain substantival at some level even if space-time points are given up, and that the program of section 10 cannot succeed.

The more resolute absolutist will want to dig in his heels at an earlier juncture. If he wants to preserve the possibility of determinism, he will have to agree with Leibniz that Leibniz-equivalent substantivalist models correspond to the same reality. But he may wish to claim that this is so not because the equivalent models provide different descriptions of a reality devoid of space-time points but rather because the identity of the points is determined by their metrical properties and relations.[20] Both sides in the absolute-relational debate complain that the devil quotes scripture. The antisubstantivalist described in section 9 quoted Einstein's remark that space-time "is a structural quality of the field" and took it to mean that space-time points exist in only a secondhand representational sense. Now the absolutist can also quote Einstein's remark, giving it the different reading that space-time points exist but have their identities settled by the structural roles they play in the *g*-field. Although my sympathies generally lie with the absolutist side of the absolute-relational debate, I must confess a number of qualms about this resolute absolutism, the most serious being that I do not see how it can be supported by a defensible account of identity and individuation. The point will be developed in the context of a discussion of the metaphysics of predication.

12 Predication and Identity

Nothing is so familiar and straightforward in its application as predication and yet so baffling in its theoretical underpinnings. What helps to make the everyday attribution of redness to individual *a* and greenness to individual *b* a straightforward matter is the assumption that *a* and *b* have already been picked out as distinct objects—an assumption that is often unproblematic in concrete applications. But to carry the assumption to its logical extreme of bare particulars—items that differ *solo numero*—is to invite a conundrum, namely, what is the principle of individuation of such items? It was precisely to avoid such unsolvable puzzles that Leibniz insisted on his principle of identity of indiscernibles.

$$(P)[P(a) \leftrightarrow P(b)] \rightarrow a = b \qquad \text{(PIdIn)}$$

And yet I believe that by asking how it is to be enforced, we can see that a kind of particularism, semibare if not entirely nude, is entailed by both traditional relationism as well as traditional substantivalism. The particularism I have in mind asserts that PIdIn will not be true in general unless the quantifier ranges over properties that are nonqualitative in the sense that they refer to particular individuals. I have no general characterization of the qualitative/nonqualitative distinction, but examples should suffice for purposes of the argument. Consider two kinds of interpretations of $P(x)$:

$P(x)$ means that	
$x = c$	x is red
$x \neq c$	x is not red
x is ten feet from c	there is a y 10 feet from x

where c names a particular individual. Under the interpretations on the left-hand list, but not the right-hand list, $P(x)$ refers to particular individuals.[21]

I take it as obvious that traditional space substantivalism entails the form of particularism in question. Let a and b denote space points, and suppose that space is homogeneous and isotropic and that it is devoid of bodies. Then to make PIdIn hold, it is necessary either to resort to items like those in the left-hand list or to postulate that the points are distinguished by, say, different hues of a radiation they give off. Traditional absolutists never thought that they had to countenance such emanations.

To explore the implications of relationism, suppose that a and b denote bodies, that these are the only bodies that exist, and that they have exactly the same size, shape, composition, etc. The absolutist can say that a and b are distinguished by their properties of location in the container space. By (R3) of section 1.5 the relationist is barred from using primitive monadic properties of spatial location, and by (R2), properties of spatial location do not derive from the relation of the bodies to a spatial substratum. And to assert that there must be some purely qualitative, nonspatial properties of a and b that ground the difference between them is to make an assertion that is unsupported by anything that physical investigations have been able to uncover. Clarke taxed Leibniz with just these considerations when he claimed that "there is no impossibility for God to make two drops of water exactly alike" and that "two things, by being exactly alike, do not cease to be two" (Alexander 1984, p. 46). Leibniz was forced to concede, "When I deny that there are two drops of water perfectly alike, or any two bodies indiscernible from each other; I don't say, 'tis absolutely impossible to suppose them; but it is a thing contrary to the divine wisdom, and which consequently does not exist" (p. 62). The concession of logical possibility already grants Clarke's point. Nor is it evident why the existence of two qualitatively indistinguishable water droplets is supposed to be contrary to divine wisdom, if by that it is meant contrary to PSR.[22]

The opposite of individuals that differ *solo numero* seems no less absurd. To say that an individual is just a collection of its properties is to utter nonsense if for no other reason than that predication assumes an object to bear the property, as Leibniz himself insisted. And yet there is a sense in which Leibniz's notion of the complete concept of an individual captures a form of the idea that an individual is a bundle of properties. Namely, the principle of the identity of indiscernibles and its converse, the principle of indiscernibility of identicals,

$$a = b \rightarrow (P)[P(a) \leftrightarrow P(b)], \qquad \text{(PInId)}$$

are both to be applied across possible worlds as well as within a world; that is, every property of an individual is an essential property, and the individual just is the thing that satisfies the list of properties constituting its essence. This account of individuation does not succeed in capturing a workable version of the traditional idea of individuals as bundles of properties since as already noted, PIdIn is confounded unless the bundle includes, as it apparently must, properties that refer to particular individuals.

Another way to try to implement the idea of individuals as bundles of properties is to try to settle matters of identity and individuation not piecemeal but collectively for an entire system of entities. On this holistic understanding of identity, PIdIn is to be applied not within a world or across worlds to individuals in the usual sense but rather to entire worlds, in which case the principle asserts that isomorphic worlds are identical. The individuals in a world are then just those entities that play such and such a structural role in the overall scheme of things. Alternatively, it might be held that the identity of various types of individuals is determined by the structural role the individuals of the given type play with respect to some circumscribed list of properties. I take it that this is the view of the resolute substantivalist of section 10 who finds support in Einstein's dictum that space-time "is a structural quality of the [g-]field," which, on the preferred reading, means that the identity of space-time points is fixed by the structural role the points play in the metric field. Metaphysics as well as politics makes strange bedfellows. Here we have the spectacle of a space-time substantivalist trying to defend his position by appealing to a Leibnizian theory of identity. The manifold substantivalist can also buy into this structural-role theory of identity, only for him the relevant structural role is played out at the level of topology and differential structure.

What is common to all these structuralist views is the notion that identity follows isomorphism, the differences being whether the isomorphism has to be total or partial, and if partial, with respect to what set of properties. As already remarked in chapter 6, this common core is in general incoherent if "identity" means literal identity and if isomorphism is not unique. If $\psi_1 : W \rightarrow W'$ and $\psi_2 : W \rightarrow W'$ are relevant isomorphisms, total or partial as the view of identity requires, and if i is an individual of W, it follows that

i is identical with $\psi_1(i)$ and with $\psi_2(i)$. And so by transitivity of identity, $\psi_1(i) = \psi_2(i)$, which gives a contradiction if ψ_1 and ψ_2 are distinct. Perhaps literal identity follows ψ_1 but not ψ_2. But since ψ_2 is an isomorphism, to fail to identify the corresponding individuals of ψ_2 is to back away from the idea that identity and individuation are settled by the structural roles played by the individuals. And it raises the problem of which isomorphisms determine identity and which do not. The intractability of this problem is, I suggest, a strong clue that the view that generates it is badly amiss.

One could try to escape these difficulties by saying of space-time points what has been said of the natural numbers, namely, that they are abstract rather than concrete objects in that they are to be identified with an order type. But this escape route robs space-time points of much of their substantiality and thus renders obscure the meaning of physical determinism understood, as the substantivalist would have it, as a doctrine about the uniqueness of the unfolding of events at space-time locations.

I conclude that the most straightforward ways of trying to understand the resolute absolutist position of section 10 are indefensible.

13 Essentialism

The discussion of section 11 and the more sophisticated arguments of Adams (1979) indicate that individuals possess a primitive nonqualitative *haecceitas* ("thisness"). It does not follow automatically that transworld identity of individuals is primitive, nor that there cannot be any qualitative necessary conditions for the thisness of an individual. I begin this section by assuming not only that these things do not follow but also that they are not true. In a provocative and learned paper that confronts Einstein with Aristotle, Tim Maudlin (1988) argues that metrical properties are essential to space-time points (a space-time point wouldn't be the very point it is if it lacked the metrical properties it actually has) and that therefore, Einstein's hole construction cannot be used to saddle the space-time substantivalist with indeterminism. In the language of the preceding section, this essentialism applies within a world and across worlds a restricted version of PInId, $a = b \rightarrow (P)[P(a) \leftrightarrow P(b)]$, in which the property variable ranges over metrical properties, but it leaves in abeyance intra- and transworld applications of PIdIn and thereby avoids the difficulties with the structural-role theory of identity. Since Maudlin's views on the details of the application of essentialism to space-time substantivalism are still evolving, it should be understood that the form of essentialism I criticize here may bear only a distant relation to the one he now wishes to advocate.

Let us temporarily set aside qualms about how the essential/accidental cut is to be drawn in order to inquire in more detail how metric essentialism is supposed to save determinism. For theories like GTR, manifold substantivalism apparently leads to automatic violations of

D For dynamically possible models, sameness before $t = 0$ implies sameness after $t = 0$.

There are three ways to prevent violations of (D) from turning into genuine violations of ontological determinism. The first is to claim that the theory is incomplete. Because the theory *under*describes the world, its dynamically possible models correspond one-many to physically possible worlds. And consequently, the splitting of models involved in the failure of (D) need not correspond to a splitting of physically possible worlds. The second way is to claim that theory *over*describes rather than underdescribes the world, that its dynamically possible models correspond many-one to physically possible worlds. This is the way of the relationist. The third way is to say that the theory *mis*describes, that some of its dynamically possible models do not correspond to physically possible worlds. This is the way of the metrical essentialist. The hole construction generates innumerably many models, each of which is intended to describes a physically possible world. According to metric essentialism, however, the intention fails save in at most one instance; indeed, all those models, except the one corresponding to the actual world, fail to describe a logically possible world, since they ascribe to some space-time points metrical properties that are contrary to the essence of the points.

Saving ontological determinism in the face of a violation of (D) requires more than bald assertion. If one claims that the failure of (D) indicates only that the theory is incomplete, then one is under obligation to show how to supplement the theory so as to restore (D). In the case of the hole construction I argued that the relevant supplementation is not forthcoming, for after additional variables are adjoined to the theory, the hole construction can be cranked up anew. Next, if one claims that the failure of (D) indicates only that the theory overdescribes, then one is under obligation to show, without loss of empirical content, how the theory can be purged of its descriptive fluff. Section 9 described how one such purging might go, but that purging went in a relationist direction. And finally, if one claims that the failure of (D) indicates only that the theory mistakenly reads physical possibility for logical impossibility, then one is under obligation to show how the theory can be cured of its mistakes.

The last challenge can be raised various ways. From the point of view of orthodox space-time theories, it seems unfortunate to try to put metrical properties on the essential side of the essential/accidental cut in the context of GTR, where the space-time metric is a dynamic field. As explained in chapter 8, fields are not properties of an undressed set of space-time points but rather properties of the manifold M, which implies that fields are properties jointly of the points and of their topological and differential properties. (It is just this fact that was taken advantage of in section 9 above to get rid of space-time points. The trick was to regard functions from space-time points to $\mathbb{R}$ as objects in their own right and then to characterize vector fields, for example, as derivations of these objects.) Thus, before the second-order field properties can be specified, first-order predications must already be in place, which presupposes that the identity and individuation of the points has been settled prior to the application of the field properties. By claiming that the identity of points depends upon the g-field, the metric essentialist is claiming that the differential geometer literally does not know what he is talking about. The substantivalist wants to appeal to the fact that the best available scientific theories quantify over space-time points as a reason for believing in the existence of space-time points. For the metric essentialist to make such an appeal is awkward, since those same theories characterize fields in a way the essentialist regards as embodying a false theory of identity. It then seems fair to ask what alternative characterization of fields the essentialist proposes to give. The only alternative I am aware of is the one sketched in section 9, but it points in a direction opposite to the one in which the essentialist-substantivalist wants to go.

Leaving aside the *technicalia* of the mathematics of field theory, there remains the fact that metrical essentialism must resort to unnatural contortions to explain the most striking feature of Einstein's GTR: the dynamic character of the space-time metric. The most straightforward way to say what this feature means is to assert, for example, that if some extra mass were brought close to some point, then the curvature at that very point would be different. But the metrical essentialist views such assertions as literally self-contradictory.[23] In addition, the metrical essentialist must resort to contortions in making sense of Witten's (1988a, 1988b) topological theory of quantum gravity, which views the space-time metric as arising through symmetry breaking.[24]

Finally, one can wonder how metrical essentialism saves determinism against the threat posed by the hole-construction argument without seriously weakening if not entirely trivializing it. Suppose that Einstein's field equations were weakened so as to allow as solutions M, g_1, T_1 and M, g_2, T_2 such that before $t = 0$, the g- and the T-fields can be matched up by a diffeomorphism, but after $t = 0$, no diffeomorphism matches them. Assuming the theory to be complete, this would generally be regarded as a genuine failure of determinism. But by metrical essentialism, if M, g_1, T_1 corresponds to the actual world, then M, g_2, T_2 does not correspond to a physically possible world or indeed to a logically possible world; hence, contrary to intuition, the bifurcation of the models of the theory does not indicate a failure of ontological determinism. Metrical essentialism must once again resort to special contortions to explain how determinism can fail. I do not say that the contortions cannot succeed, but I do say that together with the other contortions, they do make metrical essentialism unattractive as an answer to the hole problem.

The committed essentialist need not lose heart in the face of the problems surveyed above, for some of what he wants to say can be said under the protective mantle of counterpart theory.

14 Counterpart Theory

The counterpart theorist shares with the metrical essentialist the conviction that the underdetermination that the hole construction exhibits in models of GTR does not translate into an ontological indeterminism, because the theory systematically misdescribes physical and logical possibilities. Before turning to details, I want to register some general qualms both about the motivation for counterpart theory and about its application to the determinism problem.

While I have nothing new to add to the literature on transworld identity, a few points should nevertheless be stated for the record. First, if possible worlds are not to be thought of as distant planets to be discovered by use of a metaphysical telescope focused into the outer space of possibilities, but as Kripke would have it, possible worlds are stipulated, then much of the worry about transworld identification of individuals disappears. Second, one could force out the result that an individual exists in only one possible world by insisting with Leibniz that every property of an individual is essential to that individual. Such a view elevates W. C. Fields's fatal glass of beer to a ridiculous extreme: if I had had one more (or one less) glass of beer yesterday, then I would not be the very person I am. Such talk can only be read as a very deviant version of English. Third, as Nathan Salmon (1981) has shown, it is no good to claim that counterpart theory is needed because identity is a vague relation. If it is supposed to be indeterminate whether or not a is identical with b, then it is determinate that the pair (a, b) is not identical with (a, a) or with (b, b), and therefore it is determinate after all that $a \neq b$.[25] Finally, I note that the standard mathematics of field theory does not encourage the notion that there is some difficulty in identifying points of the space-time manifold across different physical situations; indeed, as I have been emphasizing, just the opposite is true.

A potential disadvantage of the application of counterpart theory to the hole problem is that a counterpart formulation of determinism would seem to take away much of the sting or, if you prefer, the excitement of indeterminism. By way of analogy, recall the well-known problem Leibniz had in explaining the basis of human freedom. Leibniz wanted to say that when Adam chose to eat the forbidden fruit, he was free to sin or not to sin. But for Leibniz there is a problem in trying to establish what is commonly regarded as a necessary condition for freedom, namely, that Adam could have done otherwise. For on Leibniz's account of personal identity, Adam would not be the very person he is if he lacked any predicate that belongs to his complete concept. Thus, 'Adam might not have sinned' cannot be true if it is counted as true just in case there is a possible world in which Adam did not sin; for that Adam look-alike who did not partake of the forbidden fruit in such and such possible world is not Adam. Of course, the semantics of subjunctive and counterfactual conditionals can be done in terms of counterpart theory, and 'Adam might not have sinned' can be counted as true just in case there is a possible world in which some appropriate counterpart of Adam does not bite the apple. But how does pointing out the actions of a *counterfeit* Adam help to establish the freedom of the *actual* Adam any more than pointing out the actions of a saint who never beheld the fruit? In a similar way, how does it help to establish the "openness" of the future of the actual world to point out that there are possible worlds that resemble the actual world in the past but not in the future if those worlds cannot be identical with the actual world for past times? Of course, as far as the substantivalist is concerned, the force of this rhetorical question is diminished if counterpart theory succeeds in reconciling determinism and substantivalism in the presence of the hole construction.

How then does counterpart theory help? Consider again the manner in which the hole construction produces an apparent example of how GTR, even with its models restricted to those possessing Cauchy surfaces, is nondeterministic. One starts with a dynamically possible model $\langle M, g, T \rangle$ and applies a hole diffeomorphism to produce another dynamically possible model $\langle M, g', T' \rangle$ that exhibits the same g- and T-states at all points of M on or before some Cauchy slice but that exhibits different states at future points. Counterpart theorists might try to save ontological determinism by following the metrical essentialists in denying that both models correspond to possible worlds. But a more plausible and natural move for the counterpart theorist is to claim that while both models correspond to possible worlds, GTR is misleading when it suggests that the same space-time points inhabit both worlds; the theory gives the same names to the points of both worlds, but strictly speaking there is no transworld identity of space-time points, so that the hole construction does not produce models where identically the same points carry the same fields in the past but not in the future. (To reflect this point of view, let us agree to write $\langle M', g', T' \rangle$ instead of $\langle M, g', T' \rangle$. This is somewhat awkward in that it appears to abandon the active form of general covariance, which demands that if $\langle M, g, T \rangle$ is a dynamically possible model, then so is $\langle M, d*g, d*T \rangle$. Such awkwardness underscores the need for the counterpart theorist to face the challenge of providing a modified apparatus for treating fields that continues to use space-time points or regions as basic objects of predication but that does not involve the systematic misdescription of physical possibilities that he claims to detect in manifold substantivalism.) While this move means that the hole construction ceases to produce an immediate conflict between determinism and substantivalism, the ultimate fate of determinism can only be discerned after the counterpart theorist has told us more about how he proposes to understand the doctrine of determinism.

The hole construction loses its power to threaten the deterministic character of theories like GTR if our criterion of determinism is not (D) but (D′).

D′ For dynamically possible models, Leibniz equivalence before $t = 0$ (in the sense of the existence of a partial diffeomorphism that matches up the field values for $t \leq 0$) implies Leibniz equivalence for all times (in the sense of the existence of a global diffeomorphism that extends the partial diffeomorphism for $t \leq 0$ and matches up the field values for all times).

The issue is whether the substantivalist can avail himself of this escape route. The relationist, of course, has a sound motivation for adopting (D′), since for him Leibniz-equivalent models correspond to the same point-free physical reality and therefore cannot be used against ontological determinism. The substantivalist can also try to motivate (D′) with the claim that Leibniz-equivalent models describe literally identical worlds populated with the same space-time points, but this claim seems to involve the incoherent structural-role theory of identity.

The counterpart theorist can try to motivate a definition analogous to (D′) in the following way. Suppose that for some pair of dynamically possible models there is a way of matching up the space-time points on or before $t = 0$ so that the matching points are counterparts and so that the counterparts have corresponding field values.[26] But suppose that there is no extension of this matching to all the space-time points so that the matching future points are also counterparts with corresponding field values. The combination of these two suppositions is a violation of determinism, and ruling out such a combination gives a definition of determinism with the structure of (D′). Call this the weak counterpart conception of determinism (WCCD). Notice that by the lights of WCCD, the general-relativistic worlds are deterministic on a purely topological-differential version of the counterpart relation. That is, if $d: M \rightarrow M'$ is a diffeomorphism and we take the corresponding points under d to be counterparts, then the hole construction fails to disturb determinism under WCCD.

But WCCD is inadequate, for the motivation leading up to this definition neglects a second way in which determinism can fail. For suppose that there is some extension of the matching of past points such that the matching future points are counterparts but do not have corresponding field values. Then it would seem that by the lights of counterpart theory the laws of nature allow for a branching in the temporal evolution of the world. Adding a clause to WCCD to rule out this possibility produces the strong counterpart conception of determinism (SCCD). But by SCCD, determinism fails in general-relativistic worlds if the topological-differential conception of counterpart is adopted, as the hole construction shows.

To protect determinism against the hole construction, the counterpart theorist apparently has to take a page from the metrical essentialist and resort to a more stringent counterpart relation involving alikeness of metric properties. Here is an initial try. For space-time M, g and M', g' say that $p \in M$ and $p' \in M'$ are counterparts just in case they share the same metrical properties in that there are isometric neighborhoods of p and p'. This is not a stringent enough notion of counterpart, since the hole construction still produces a violation of SCCD.[27] Stipulating that the counterparthood of p and p' requires that M, g and M', g' are globally isometric does suffice to make SCCD proof against the hole construction, but it also trivializes determinism.

As suggested by Butterfield (1989), perhaps the best course for the counterpart theorist is to apply the counterpart relation not to individual space-time points but to regions. Consider regions $R \subset M$ and $R' \subset M'$ respectively of the space-times M, g and M', g'. If there is a diffeomorphism $d: R \to R'$ that is an isometry, call R and R' counterparts under d. Call a region a *past* (respectively, *future*) if it consists of all points lying on or below (respectively, above) a global spacelike hypersurface. Then determinism for GTR can be taken to mean that for any pair of dynamically possible models, if the pasts belonging to some Cauchy surfaces of those models are counterparts under some diffeomorphism d that matches up the values of nonmetric fields at past points, then d is extendible to a d' under which the futures are counterparts and the future points have matching nonmetric field values. No additional content is gained by adding that any other extension d'' that makes the futures counterparts also matches up the nonmetric field values.

This latest definition of determinism succeeds in making determinism immune to the hole construction by loading the elastic notion of counterpart with much of the content of the doctrine of determinism; indeed, in the matter-free case the latest counterpart notion bears the full weight of determinism! In addition, this stretching of the elastic also removes some of substantiality from space-time points. If we are to think of these entities as very small, momentary, and immaterial physical objects, then the counterpart relation should be applicable to them individually, which is just what the latest definition denies. And any attempt to make the counterpart relation so applicable seems to either run afoul of the hole construction or to trivialize determinism.

Finally, some of the objections that applied to metrical essentialism also apply to the latest counterpart approach to determinism. In particular, if metrical properties are essential to the counterpart relation as applied to space-time points or regions, then a counterpart reading of subjunctive conditionals would seem to block the most direct way of expressing the dynamic character of the space-time metric by means of such assertions as 'If extra mass were brought into this region, its curvature would be different'. For there is no possible world in which a counterpart of this region has a different curvature, since different curvature implies a different metric, which in turns implies that the regions are not counterparts. Of course, we know that there are many different counterpart relations, and perhaps one that cues to metrial properties is appropriate for discussing determinism, while one that cues to nonmetrical properties is appropriate to analyzing subjunctive conditionals like the one in question. But since subjunctive and counterfactual conditionals, laws of nature, and determinism form a tight circle of concepts, this lack of uniformity cries out for some principled motivation to guide the stretching of the elastic counterpart notion. And if the lack of uniformity means that the doctrine of determinism does not issue in the expected counterfactuals of the form 'If the past state had been ..., then the present and futures states would be ...', it speaks against the counterpart solution to the hole construction.

15 Conclusion

In this chapter I have tried to show how the considerations raised in Leibniz's famous argument against substantivalism map onto a set of foundation problems in GTR that were present from its very inception. I have also tried to show how these problems can in turn be used to enrich and enliven the philosophical discussion of the issue of relationism versus substantivalism. In previous chapters I have argued that Leibniz's original version of the PSR-PIdIn argument was polemically ineffective, as are the reconstructed versions offered by Friedman and Teller. I have claimed here that the causal version of the PSR argument, as embodied in Einstein's hole construction, is polemically effective at least to the extent that the substantivalist who wishes to secure the possibility of determinism is forced to abandon one form of space-time substantivalism, what I have called manifold substantivalism. There may, of course, be other defensible versions of substantivalism that escape the hole construction, but our initial survey of the possibilities was not encouraging. The way thus seems open for the relationist to claim that the possibility of determinism is secured, because Leibniz-equivalent substantivalist models correspond to the same physical reality, for in using the fictional entities of space-time points, these models provide one-many representations of a point-free reality. The claim requires substantiation. As illustrated in section 9, one can render the reality underlying the Leibniz-equivalence classes free of space-time points, but the rendering suggested there was substantival at a deeper level.

My own tentative conclusion from this unsatisfactory situation is that when the smoke of battle finally clears, what will emerge is a conception of space-time that fits neither traditional relationism nor traditional substantivalism. At present we can see only dimly if at all the outlines the third alternative might take. But I hope to have identified the considerations we need to pursue in trying to give it a more definite form. And I hope that even those readers who do not accept the morals I draw for the absolute-relational debate will nevertheless agree that the hole construction and the catalog of reactions to it serve both to reveal a previously unappreciated richness to the doctrines of determinism and space-time substantivalism and to link these doctrines in a deeper way to issues in the philosophy of science and in metaphysics.[28]

Notes

Introduction

1. Occasionally, articles on the absolute-relational dispute appear in the scientific literature, for example, Erlichson 1967 and Heller and Staruszkiewicz 1975. Of more importance here is the work of Barbour and Bertotti, discussed in chapter 5.

Chapter 1

1. The rivals of the Reverend Clarke repeated the snide but not wholly inaccurate remark that he would have made a good Archbishop of Canterbury if only he had been a Christian, a reference to his Arianism (see Hall 1980). There can be no doubt that Clarke had Newton's confidence, because Newton had authorized him to do a Latin translation of the *Optics*. Koyré and Cohen (1962) argue that Newton had a hand in drafting Clarke's responses to Leibniz (but see also Hall 1980). However, there are points at which Clarke does depart from Newton's own views on space; see section 6.1 below.

2. This theme is also found in Mach's criticism of Newton; see Mach's *Science of Mechanics* and section 4.8 below.

3. Any reader who, after completing this book, believes that any of Reichenbach's theses stands without major qualifications is cordially invited to commit this book to the flames.

4. For an analysis of the doctrine that space is an emanent effect of God, see McGuire 1978.

5. In C. D. Broad's (1946) terminology, this is an assertion of the adjectival over the qualitative theory.

6. See Horwich 1978 and Teller 1987 for discussions of the property view.

7. See my 1978.

8. The first quotation is from "First Truths" (1680–1684; Loemker 1970, p. 270); the second is from the correspondence with Arnauld (1687; Loemker 1970, p. 343).

9. See McGuire (1976) and Cover and Hartz (1986).

10. From "Reply to the Thoughts on the System of Preestablished Harmony..." (1702; Loemker 1970, p. 583).

11. From a letter to de Volder (1706; Loemker 1970, p. 539).

12. I have not been able to pinpoint the beginnings of this doctrine. In the 1671 essay "Studies in the Physics and Nature of Body," Leibniz begins by apparently affirming a commonsensical view of the continuum: "*There are actually parts in a continuum.... And these are actually infinite*" (Loemker 1970, p. 139). But then he continues, "*There is no minimum in space or in a body*" because otherwise it would follow that "there are as many minima in the whole as in the part, which implies a contradiction."

13. See, for example, Malament 1977 for an elucidation and criticism of this doctrine.

14. Lucas 1984, p. 194. A very similar idea is found in Feigl 1953.

15. In the corollaries to proposition VI of book 3 of the first edition of the *Principia*, Newton maintained against Descartes that not all spaces are equally filled and, thus, that there are vacuums in nature. This claim, which was the subject of correspondence with Cotes (see Hall and Tilling 1975) is modified in the second edition, where Newton makes only the conditional assertion that "If all the solid particles of all bodies are of the same density, and cannot be rarefied without pores, then a void, space, or vacuum must be granted."

16. See Friedman 1983, chapter 6, for further discussion.

17. See Wald 1986.

18. For this reason I make no apology for neglecting, at least initially, various issues and positions taken on these issues.

Chapter 2

1. The standard summation convention on repeated indices is in effect. Latin indices run from 1 to 4; Greek indices from 1 to 3. The condition that $g^{ij}t_j = 0$ is well defined under different choices for the allowed t, for if $t \to t' = f(t)$, $df/dt > 0$, then $t'_i = t_i(df/dt)$ and $g^{ij}t'_j = 0$ if and only if $g^{ij}t_j = 0$.

2. Unless otherwise specified, the space-time manifold is assumed to be C^∞, as are vector fields and congruences of curves defining reference frames.

3. Chapter 5 will tackle the more difficult matter of relativistic rigid motion.

4. See the appendix to chapter 3 for further discussion of this notion.

5. We want ε_{ijkl} to be totally antisymmetric, i.e., $\varepsilon_{ijkl} = \varepsilon_{[ijkl]}$. This fixes the volume element up to scale factor which can be set be requiring that in a coordinate system in which $g^{ij} = \text{diag}(1,1,1,0)$ and $h_{ij} = \text{diag}(0,0,0,1)$, $\varepsilon_{ijkl} = (dx_1)_i \wedge (dx_2)_j \wedge (dx_3)_k \wedge (dx_4)_l$. John Norton has shown in a private communication that this condition is equivalent to requiring $g^{jj'}g^{kk'}g^{ll'}\varepsilon_{ijkl}\varepsilon_{i'j'k'l'} = 3!\,(dx^4)_i(dx^4)_{i'}$ and that these requirements determine ε_{ijkl} uniquely.

6. If, for example, $V^i = (\omega y, -\omega x, 0, 1)$, with ω constant (i.e., there is a uniform rotation about the z axis), $\Omega^i = (0, 0, \omega, 0)$ in a coordinate system that is inertial with respect to one of the preferred connections that define nonrotating motion.

7. See, for example, Havas 1964; Kuchar 1981; and Torretti 1983.

8. Answers to the quiz. Q1: 6 only. Q2: 5+. Q3: 5+. Q4: 5+. Q5: 5+. Q6: 3+. Q7: 2+. Q8: 2+. Q9: 1+. Q10: 5+. Here $n+$ means that the question is meaningful for the space-time n and all of those that follow.

9. Here d^* stands for the dragging along by the diffeomorphism d. The bold $\mathbf{O}_i$ denotes an object type and O_i denotes a value. See section 8.3 for more on the notion of a geometric object.

Chapter 3

1. Descartes adds the qualification "considered at rest" as a sop to ordinary language. Without this qualification it follows that "transference is reciprocal; and we cannot conceive of the body *AB* being transported from the vicinity of the body *CD* without also understanding that the body *CD* is transported from the vicinity of body *AB*.... [Therefore,] we should say that there is as much movement in the one as in the other. However, [I admit that] that would depart greatly from the usual manner of speaking" (1644, II.29).

2. Translations of the passages from Huygens were kindly provided by M. Spranzi-Zuber and J. E. McGuire.

3. On theories of meaning that take the meaning of a word or concept to be given by the associated sense impressions, there is no real distinction between the meaning claims and the epistemological claims.

4. Or at least this is what Huygens should have objected; whether he fully grasped the point is not clear from the text. For further discussion of this point, see chapter 4 below.

5. See Hempel 1965, van Fraassen 1980, and Salmon 1984 for various views on this matter.

6. See Armstrong 1983 and chapter 5 of my 1986 for a review of various positions on this issue.

7. See the appendix of this chapter for a discussion of other conceptions of dynamic symmetries.

8. See Rynasiewicz 1986 for further analysis of the relevant concepts.

9. Einstein perceived the development as moving in this direction, and the perception may have helped to lead him to STR (see Earman et al. 1983). In "Ether and the Theory of Relativity" Einstein wrote: "As to the mechanical nature of the Lorentzian ether, it may be said of it, in a somewhat playful spirit, that immobility is the only mechanical property of which it has not been deprived by H. A. Lorentz. It may be added that the whole change in the conception of the ether which the special theory of relativity has brought about consisted in taking away from the ether its last mechanical quality, namely, its immobility" (1920, pp. 10–11).

10. This and other aspects of determinism are discussed in my 1986.

Chapter 4

1. Before proceeding further the reader may wish to review the Scholium, reproduced as an appendix to chapter 1.

2. Newton intimates that he has performed the experiment ("as I have experienced"). Whether or not Newton actually carried out the experiment in the form described in the Scholium is beside the point, since the relevant phenomena are a matter of daily experience.

3. See chapter 3 above and Laymon 1978, pp. 404–407.

4. Think about it for a moment. Would you seriously entertain the possibility that the reason you get seasick is due to your motion relative to the stars? Is this notion any more plausible than astrology?

5. Cotes, not Gregory, was responsible for the new edition. Had Huygens lived to see it, he would have found no retraction. The translations of the Huygens–Leibniz correspondence used in the text are taken from Degas 1958, pp. 490–492. M. Spranzi-Zuber and J. E. McGuire kindly provided the translations of Huygens's manuscripts on absolute motion.

6. The cagey, cards-held-close-to-the-vest style of this part of the Huygens–Leibniz correspondence is so obvious that only a very strong preformed judgment can explain Reichenbach's characterization of the exchange: "Apparently, the belief that he could expect a complete understanding of his ideas from his former mathematics teacher encouraged Leibniz to be more open in his communications with Huyghens.... Discussions among like-minded scholars who are working in the same field are usually briefer in their expositions than those presentations intended for a wider audience.... The lucid style of this correspondence reveals at once a most fortunate meeting of minds" (1924, pp. 59–60).

7. As befits the importance of the topic, Huygens wrote in Latin. A German translation of some passages was given by Schouten (1920). English translations of some passages are to be found in Stein 1977 and Bernstein 1984, but no complete English translation has been published. The commentaries of Stein and Bernstein are the best available, and although I differ from each on several points, my discussion owes much to theirs.

8. Proposition 5 of Huygens's *De vi centrifuga* (1659) reads, "When a mobile moves on a circumference with the velocity it would acquire in falling from a height equal to a quarter of its diameter, its centrifugal force is equal to its gravity; in other words, it will pull on the string by which it is attached with the same force as if it were suspended by it" (Huygens 1888–1950, 16:275; trans. Degas 1958, p. 296).

9. Here I am in agreement with Bernstein (1984, p. 92).

10. The first part of Leibniz's "Specimen dynamicum" is at pains to point out that Descartes's laws of impact cannot be correct, because they violate Galilean relativity. It was Huygens who first derived the correct laws of impact for perfectly hard (i.e, elastic) bodies. His derivation heavily exploited Galilean invariance.

11. Compare this to Newton's corollary 6 to the Axioms of *Principia*: "If bodies, moved in any manner among themselves, are urged in the direction of parallel lines by equal accelerative forces, they will all continue to move among themselves, after the same manner as if they had not been urged by those forces" (1729, p. 21).

12. For further details of how classical and quantum mechanics can be done in the setting of Maxwellian space-time, see Hood 1970 and Rosen 1972.

13. Of course, Leibniz and other founders of relationism would have rejected any action-at-a-distance formulation of mechanics.

14. See, for example, Loemker 1970, p. 419, and Gerhardt 1849–1855, 6:144–147.

15. These examples are suggested by the thought experiments discussed by Newton in "De gravitatione"; see section 1 above. For further discussion of the instrumentalist move, see section 9 below.

16. Recently it has been claimed that Mach's supposed rejection of relativity theory in the preface to the second edition of the *Optics* was a fabrication of his son Ludwig; see Harré 1986, pp. 15–16. Whatever the merits of this claim, it seems clear that a negative attitude toward relativity theory flows naturally from Mach's general philosophical orientation.

17. For Mach's influence on Einstein see Holton 1973, pp. 219–259. See chapter 5 below for a discussion of the relativistic treatment of rotation.

18. Poincaré's reasoning is sophistical. He seems to be hankering here for a space-time in which rectilinear motion is well defined but rotation isn't. Can there be such a space-time?

19. This idea is found in current textbooks; e.g., Griffiths (1985, p. 82) says: "In the application of classical dynamics it is therefore essential to determine initially an inertial frame of reference."

20. Here 'second-order' and 'third-order' have their usual meanings, not Poincaré's non-standard meanings.

21. Paul Teller in a private communication. Note that some of the examples Clarke urges against Leibniz involve motion in a line.

Chapter 5

1. The flexibility afforded by Machian space-time in choosing a time parameter will come into play below.

2. As required to make $Ld\lambda$ invariant under $\lambda \to \lambda' = f(\lambda)$, $df/d\lambda > 0$.

3. That is, $(d/d\lambda)(\partial L/\partial \dot{r}) - (\partial L/\partial r) = 0$ follows directly from $(\partial L/\partial \dot{r})r - L = 0$.

4. Before switching to local time, the first term on the right hand side of (5.6) is multiplied by a factor proportional to M^3/R^2 (as would be expected from [5.4]), so that in cosmic time the total momentum $\sum_i \partial L/\partial r_i$ is not constant if R is changing.

5. The 1982 theory of Barbour and Bertotti was designed to escape this difficulty. However, this new theory seems to amount to a reworking of the approach of Zanstra (1924).

6. I will work with signature $(+ + + -)$. Since there are many excellent textbooks covering the relevant material on relativistic space-times (e.g., Hawking and Ellis 1973 and Wald 1984), I will not review any of it here, save for the topic of relativistic rigid motion, which can be found only in more inaccessible sources.

7. As follows from differentiating $V^i V_i = -1$.

8. The symbol £ denotes the Lie derivative. See the appendix to this chapter for computations using £.

9. A follow here the usual convention of using round and square brackets around indices to indicate respectively symmetrization and antisymmetrization.

10. In Minkowski space-time the stationarity of V^i implies that the magnitude of rotation is constant along the trajectories of V^i. In a curved space-time, constancy of rotation can be deduced from stationarity plus two additional assumptions, namely, that V^i is an eigenvector of T^{ij} and that Einstein's field equations are satisfied.

11. Of course, it is open to respond that this confutation shows that Born rigidity is not an adequate explication of rigid motion. But this response carries with it the responsibility to provide an alternative explication.

12. Temperature differentials are the most plausible explanation of Miller's results (see Shankland 1964 and Shankland et al. 1952.

13. This assumes that the space-time is temporally orientable.

14. See Hawking and Ellis 1973 and Malament 1985.

15. See Hawking and Ellis 1973.

16. See Adler et al. 1975, p. 443, for a detailed justification of this identification.

17. Adler et al. 1975, pp. 446–447.

18. A result due to Carter (1973) shows that under certain global restrictions, stationarity implies nonrotation; specifically, if (1) the space-time manifold M is topologically $\mathbb{R}^4$, (2) the metric is asymptotically Minkowskian, (3) Einstein's field equations hold, (4) the velocity field V^i of matter is a Killing field that is nonvanishing on M, and (5) V^i is an eigenvector of the energy-momentum tensor, then V^i is nonrotating.

19. The connection between this issue and space-time substantivalism will be taken up in chapter 9.

20. The next Weyl cites is a letter to de Volder (Gerhardt 1875–1890, 2:168–175; Loemker 1970, pp. 515–518). The main thrust of the letter is Leibniz's familiar attack on Descartes's notion that the concept of extension suffices to explain inertia and a reiteration of the claim that substance is not constituted by extension alone but requires "force."

21. The sense in which uniqueness can be expected is a little tricky to pin down (see chapter 9).

Chapter 6

1. This is probably at best a partial explanation, for as noted in chapter 4, I suspect that Leibniz was not confident of his response to Newton's bucket experiment and that this lack of confidence may have made him reluctant to raise the matter in the correspondence with Clarke.

2. See Clarke's Boyle lectures (1738, 2:527–530).

3. Newton's account is prefaced by the remark, "I am reluctant to say positively what the nature of bodies is, but I rather describe a certain kind of being similar in every way to bodies" (Hall and Hall 1962, p. 138).

4. The ambiguity is not an artifact of the English translation of the original French ("par un échange d'orient et de l'occident" [Gerhardt 1875–1890, 7:364]). The German translation available to Kant would not have suggested the ambiguity; see chapter 7 below.

5. The importance of the distinction between the causal and deterministic reading is brought out in my 1986.

6. There is also the question of whether the proper application of PIdIn is to individuals within a world. Transworld applications and the implications for substantivalism are treated in section 7 below and in chapter 9.

7. Another attempt by Friedman (1983) to state a nonverificationist version of Leibniz's argument is considered in chapter 9.

8. The correspondence represents but a small piece of an ongoing dispute, the most bitter part of which centered on the priority for the invention of the calculus (see Hall 1980).

9. Indeed, Einstein's GTR forces us to take this possibility seriously.

10. Friedman (1983) and others credit Sklar with a major new insight. I believe that Sklar is to be credited with a very clever conjuring trick. However, the trick points to a very interesting view, which will be explored in chapters 8 and 9.

11. As noted by Friedman (1983, p. 233).

12. As suggested by Friedman (1983, p. 233, n. 10).

13. I am grateful to Tim Maudlin for urging me to consider the kind of approach described in this section.

14. I am indebted to Robert Geroch for this point.

15. Newton's Scholium on absolute space is attached to definition VIII.

16. Actually what Leibniz says is that force "is of importance not only in physics and mechanics in finding the true laws of nature and rules of motion, and even in correcting many errors which have slipped into the writing of a number of able mathematicians, but also in metaphysics for the better understanding of the principles" (Loemker 1970, p. 315).

17. The translation is by Daniel Garber.

18. This absence makes one suspect the 1688 date assigned to the manuscript.

Chapter 7

1. Euler's paper appeared in the *History of the Royal Berlin Academy*, 1748. An English translation can be found in Koslow 1967.

2. The full quotation reads: "Let it be imagined that the first created thing were a human hand, then it must necessarily be either a right hand or a left hand. In order to produce the one a different action of the creative cause is necessary from that, by means of which its counterpart could be produced" (1768, p. 42).

3. There was a 1720 edition by Heinrich Kohler and a second edition in 1740.

4. Van Cleve (1987) argues that orientable, $(n + 1)$-dimensional spaces serve the same function here as do nonorientable n-dimensional spaces. I think that this contention is incorrect. First, he maintains that "a nonorientable space of n dimensions is possible only in an ambient space of $n + 1$ dimensions" (p. 45). This contention is false. One way to conceive of a nonorientable n space is to embed it in an orientable $n + 1$ space, just as one way to picture a curved n space is to embed it in a higher dimensional flat space (in general, a flat space of $n(n + 1)/2$ dimensions is needed). But nonorientable and curved spaces exist in their own right: they possess intrinsic characterizations, and there is no need to define them by means of higher-dimensional embedding spaces. Second, he takes higher-dimensional, orientable spaces to show that "there are possible spaces in which a solitary hand would be neither right nor left; hence a hand with the same internal relations as a given right hand might fail to be right, for

it might be the sole occupant of such a space. So the rightness of a hand is not entailed by its having the internal relations it does, contrary to internalism" (p. 52). But an *n*-dimensional "hand" in an $n + 1$ space is not a hand in even the most minimal sense; namely, it is not an enantiomorph (see definition below). How, then, can considerations of higher dimensions, which convert hands into "hands," show anything about hands? Perhaps the answer is that higher-dimensional spaces show that the enantiomorphism of an object depends upon the dimensionality of the space and, in that sense, is not an internal property of the object. But that sense is irrelevant to the absolute-relational controversy. If the relationist cannot give a satisfactory account of the dimensionality of space, then relationism fails even before the question of incongruent counterparts arises. If the relationist can produce a satisfactory account, then the question becomes whether *n*-enantiomorphism in an *n*-dimensional space is a matter of the internal relations of the body. I show in section 2 that the relationist can give a positive answer to this question. I also show why being right- or left-handed is not the sort of property a relationist would want to count as being internal.

5. The full quotation reads, "The region, however, to which this order of the parts is directed, is related to space outside, but not with reference to its localities, for this would be nothing else than the position of just those parts in an external relation; region is rather related to space in general as a unity, of which each extension must be regarded as a part" (1768, p. 37).

6. Such a space must be of constant curvature.

7. This definition presupposes that space is locally orientable, a presupposition assured by the restriction to spaces that are manifolds. 'Reflection' means a single reflection. The product of an even number of reflections is equivalent to a rotation and/or translation. See note 11.

8. Here I part company with Nerlich (1976, p. 35), whose definition of enantiomorphism is nonlocal.

9. Jeremy Butterfield and Michael Redhead have urged that more should be required of the relationist; namely, he should have a general relationist definition of 'enantiomorph' and should demonstrate that all and only objects satisfying this definition will satisfy the absolutist definition of enantiomorph under any appropriate absolutist representation. They are skeptical that this demand can be met.

10. Perhaps this is the place where, as Sklar (1974) suggests, the relationist must make use of *possibilia.* There are other places where *possibilia* enter the discussion; see, for example, sections 6.12 and 8.6. However, I do not think that the core of the absolute-relational controversy turns on the use or status of *possibilia.*

11. In three spatial dimensions the parity operation is $x \to -x, y \to -y, z \to -z$. In spaces of an even number of dimensions the parity operation does not correspond to mirror image reflection, since the product of an even number of reflections is equivalent to a rotation and/or translation.

12. This picture is taken from Sakurai 1964.

13. The meaning of invariance under the parity transformation needs a bit of explanation. Let ψ^{P} denote the parity image of ψ. Since in three spatial dimensions the parity operation is equivalent to mirror image reflection plus a rotation, we can write M (for mirror) in place of P. Invariance under the parity operation demands that

If $\psi_1 \to \psi_2$ by Schrödinger evolution, then Schrödinger evolution also requires that $\psi_1^{\mathrm{M}} \to \psi_2^{\mathrm{M}}$.

The transition probability from initial state ψ_{i} to final state ψ_{f} is

$$\mathrm{P}_{\mathrm{if}}(t, t_{\mathrm{o}}) = |(\psi_{\mathrm{f}}, T(t, t_{\mathrm{o}})\psi_{\mathrm{i}})|^2,$$

where $T(t, t_{\mathrm{o}})$ is the time evolution operator and (,) is the inner product on the Hilbert space of states. The demand of parity invariance can be shown to be equivalent to $P_{if} = P_{if}^{M}$, i.e., the probabilities for the process and its mirror image are the same:

$$|(\psi_f, T(t, t_o)\psi_i)|^2 = |(\psi_f^M, T(t, t_o)\psi_i^M)|^2.$$

Now if $\psi_i = \psi_i^M$, as in the experiment I describe, then

$$|(\psi_f, T(t, t_o)\psi_i)|^2 = |(\psi_f^M, T(t, t_o)\psi_i)|^2,$$

i.e. the mirror-image final outcomes ψ_f and ψ_f^M should be equally likely. But since the probabilities (as evidenced by the statistics) are different in the experiment, parity is violated. Using the type of experiment in which $\psi_i = \psi_i^M$ helps to alleviate problems in communicating to the inhabitants of a distant planet the difference between right and left (in the sense of Which is which?). Compare this to Bennett 1970.

14. There is a possible but measure-zero set of cases in which the relative frequencies do not conform to the propensity probabilities.

15. The sense of universality appropriate here is not captured by conditions on the syntactic form of the law statement (e.g., that laws are written in universally quantified form or that their mathematical expression is generally covariant). Rather, the requirement is that the law be invariant under space-time translations. Such invariance conditions are best expressed in terms of semantic-model theory. See Rynasiewicz 1986.

16. When we take into account space-time and CPT invariance (that is, invariance under the combined operations of charge conjugation, parity, and time reversal), a more complicated conclusion emerges. See Earman 1971.

17. Although something of a straw man, the conservative interpretation is nevertheless a useful straw man. It is not too far removed from the reading Broad (1978) gives.

18. See also Kant 1790s, where it is claimed that Leibniz cannot account for the fact that 'Space is three dimensional' is "an apodictic a priori proposition" (p. 89).

19. In both the 1768 essay and the *Prolegomena* Kant uses the same German word 'inner', which is variously translated as 'inner' and 'internal'.

20. Broad attempts to preserve a conservative reading of the *Prolegomena* by taking Kant to be arguing, first, that incongruent counterparts show that space is absolute and, second, that absolute space has a property that is incompatible with its being a thing in itself: "In absolute space the existence and nature of every part would be dependent upon the existence and nature of the whole" (Broad 1978, p. 41). But Broad's reconstruction is a very strained reading of Kant's announced claim that the "paradox" of incongruent counterparts itself leads one to suspect that "the reduction of space and time to mere forms of sensuous intuition may perhaps be well founded" (1770, p. 29). In addition, the *Prolegomena* does not contain the quotation Broad attributes to Kant but rather the assertion "That is to say, the part is possible only through the whole, which is never the case with things in themselves" (1783, p. 30). The phrase 'that is to say' indicates that Kant was not giving a separate argument but only emphasizing the previous line, "Space is the form of external intuition of this sensibility."

21. Starting from some of the observations made above, Harper (1989) has written a thoughtful study of Kant's use of incongruent counterparts. I will not spoil the reader's pleasure of discovering this piece for himself.

Chapter 8

1. Field defines *reductive relationism* to be the view that space-time points are set-theoretic constructions out of physical objects and their parts, while *eliminative relationism* holds that

it is illegitimate to quantify over space-time points at all (Field 1983, p. 34). Although useful and important, this distinction will not feature in the discussion below.

2. For a collection of reprints of relevant articles, see Kerner 1972.

3. For simplicity I will deal with C^∞ manifolds. A "smooth map" is thus a C^∞ map. The definitions given below are adapted from Hawking and Ellis 1973 and Wald 1984, to which the reader is referred for further details.

4. M is Hausdorff just in case for any distinct $p, q \in M$, there exist open sets $U, V \subset M$ such that $U \cap V = \varnothing$ and $p \in U$ and $q \in V$. M is paracompact if every open cover of M has a locally finite refinement.

5. The standard treatment of geometric objects goes back to Schouten and Haantjes 1936. The definition given here comes from Trautman 1965.

6. Salvioli (1972), for example, uses the fact, assumed above, that geometric object fields are well-defined under the operation of dragging along by a diffeomorphism d; namely, if $F:(p, x^i) \to (F_1, F_2, \ldots, F_N)$, then $d * F:(d(p), d * x^i) \to (F_1, F_2, \ldots, F_N)$, where $d * x^i(p) = x^i(d^{-1}(p))$.

7. Once again, it is assumed that M is Hausdorff, paracompact, and without boundaries. That M carries a Lorentz signature metric implies the additional restriction that M must be noncompact or else have an Euler characteristic of zero.

8. Ideally, the first part of the challenge should be met by means of conditions formulated in terms of the relationist's models of reality. Since the relationist has not supplied such as yet, we have to work in terms of the substantivalist models.

9. This definition is due to Geroch (1977).

10. See the discussion in section 1.6.

11. See the discussion in section 6.6.

12. See Friedman 1983, p. 221, and, *mea culpa*, my 1970.

13. Friedman thinks that there is an objection to substantivalism that is independent of observationality; see section 8 below.

14. Here I am echoing the points made by Nerlich (1976, chapter 2).

15. This condition neglects the fact that even if we fix on Minkowski space-time, there are, according to the substantivalist, many Minkowski space-times according to how the Minkowski metric is placed on $\mathbb{R}^4$.

16. Penrose (1971) and Kaplunovsky and Weinstein (1985) provide programmatic sketches of such creative alternatives.

17. On this matter see Mundy 1986.

18. It was the struggle against such defeat that led Huygens, Leibniz, and others to such extreme lengths in trying to account for rotation (see chapter 4).

19. Field (1985) has argued in addition that the relationist faces difficulties in accounting for physical quantities, a charge originally made by Clarke in the Leibniz–Clarke correspondence. For a response, see Mundy 1987.

Chapter 9

1. See Einstein 1914a, 1914b and Einstein and Grossmann 1913, 1914.

2. For a detailed analysis of the various versions of Einstein's hole argument, see Norton 1987. For a new interpretation of Einstein's use of coordinate terminology, see Norton 1988. The reader should also consult Stachel 1986.

3. For a discussion of this matter, see chapter 3 of my 1986.

4. For a more precise version of this construction, see Earman and Norton 1987.

5. Recall relationist thesis (R3) from section 1.5.

6. I have taken the liberty of substituting my numbering of equations for Møller's.

7. Møller cites Hilbert 1915, but the correct reference is Hilbert 1917.

8. Hilbert cites a 1914 version of Einstein's hole argument; see Norton 1987. Don Howard and John Norton (private communication) have turned up evidence that the circle closes in that Hilbert's response to the hole construction was communicated to Einstein by a Göttingen colleague of Hilbert's. How Einstein's understanding of the hole argument was influenced by this correspondence is still a matter of conjecture.

9. I have taken the liberty of altering the notation to conform to mine.

10. I am grateful to Arthur Fine for emphasizing this point to me.

11. No. 9 372, Einstein Archive, Princeton University. The English translation is from Norton 1987.

12. This was suggested to me by Michael Redhead.

13. As suggested, though not advocated, by Jeremy Butterfield in a private communication.

14. A completely regular space X is such that for every $p \in S$ and every open neighborhood U of p, there is a continuous real valued function f on S such that $f(p) = 0$ for all $p \in U$ and $f(x) = 1$ for all $x \in S - U$. A completely regular space X is first-countable if for every $p \in S$, there is a countable basis for the open sets containing p.

15. Geroch (1972) uses the term "Einstein algebra."

16. As we have already seen, $C_o(M)$ and $C_o^*(M)$ determine the topology of M. Once the topology is fixed, the ring $C_\infty(M)$ determines the differential structure; see Nomizu 1956.

17. In more detail, let $\mathscr{R}(\mathscr{L})$ stand for the set of all space-time realizations of $\mathscr{L}$. For each member of $\mathscr{R}(\mathscr{L})$, choose a Cauchy surface, and let $\mathscr{R}(\mathscr{L})|_{\mathscr{S}}$ stand for the restriction of the members of $\mathscr{R}(\mathscr{L})$ to those times up to and including the instants on the chosen Cauchy surfaces. The demand of minimal Laplacian determinism is then that if $\mathscr{S}$ and $\mathscr{S}'$ are such that $\mathscr{R}(\mathscr{L})|_{\mathscr{S}} = \mathscr{R}(\mathscr{L}')|_{\mathscr{S}'}$ then $\mathscr{R}(\mathscr{L}) = \mathscr{R}(\mathscr{L}')$.

18. This assumes that the space-time is temporally orientable.

19. See chapter 10 of my 1986 and chapter 7 of Horwich 1987.

20. Newton might be read as propounding such a view in "De gravitatione" when he wrote:

The parts of space derive their character from their positions, so that if any two could change their positions, they would change their character at the same time and each would be converted numerically into the other. The parts of duration and space are only understood to be the same as they really are because of their mutual order and position; nor do they have any hint of individuality apart from that order and position which consequently cannot be altered. (p. 136)

Note that this passage is designed to support Newton's assertion that "the parts of space are motionless." Now suppose that Newton had lived to participate in the development of theories in which the metric of space-time admits no rigid motion, so that the points of space, however defined, are not motionless; i.e., they change their distances with respect to one another. Would Newton have then wanted to say that as a result, the identity of space points changes with time? I doubt it. Imagine also that Newton came to realize that the metric of space or space-time is different in different physically possible situations. Combine that realization with Newton's treatment of matter in terms of field-theoretic properties of space (see chapter 6).

Since the metric becomes just another dynamical field, one could predict that Newton would have adopted what I call manifold substantivalism. This position is consistent with his statement that the points of space do not have an individuality apart from their order and position if this phrase is interpreted in terms of topological and differential properties rather than in terms of metrical properties.

21. For a more careful discussion of this distinction, see Adams 1979.

22. Hacking (1975) maintains that if we are clever enough, we can always redescribe the situation in such a way that Leibniz's interpretation of the PIdIn is preserved; e.g., we can describe it in such a way that there is only one rain drop instead of two. This is too clever by half. If there are no constraints on our descriptions, then Hacking's claim is correct but uninteresting. On the other hand, if we accept the constraints of the best available scientific theories, whether Newtonian, relativistic, or quantum, then the redescription ploy is ruled out, since those theories all allow for the existence of states falsifying the Leibnizian alternative.

23. Michael Redhead (private communication) has urged this point.

24. I am grateful to Robert Weingard for bringing this point to my attention.

25. In fairness, it has to be acknowledged that there are interesting arguments on the other side; see Lewis 1986.

26. In Butterfield's (1989) version of counterpart theory for space-time points, the counterpart definition cues to all fields, nonmetrical as well as metrical, which makes the second conjunct of the supposition redundant. This view, however, does not affect the difficulties discussed in the next two paragraphs if we consider empty-space solutions to GTR. This brief section does not do justice to Butterfield's rich and ingenious discussion, which, in my opinion, offers the substantivalist the best method for dealing with the hole construction.

27. In fact, there will be dynamically possible models $\langle M, g, T \rangle$ and $\langle M', g', T' \rangle$ with the following properties. T and T' vanish identically. M, g and M', g' have Cauchy surfaces S and S' respectively such that there is a diffeomorphism ϕ from the past of S to the past of S' that is an isometry. Further, there is an extension of ϕ to $\tilde{\phi}$ that maps M onto M' in such a way that for any $p \in M$ to the future of S, p and $p' = \tilde{\phi}(p)$ are counterparts in the present sense; that is, there are neighborhoods $N(p)$ and $N'(p')$ and a diffeomorphism $\psi : N \rightarrow N'$ that is an isometry. And yet $\tilde{\phi}$ is not an isometry for all future points. The counterpart theorist may respond that we need a definition of determinism that requires not only that the matching points of $\tilde{\phi}$ be counterparts but also that they be counterparts under $\tilde{\phi}$ itself. This has the effect of changing the focus of the counterpart relation from individual points to regions, as is discussed below.

28. The issues raised in this chapter have already touched off a lively debate. For a sampling of opinions, the reader is referred to the articles by J. Butterfield, T. Maudlin, J. Norton, and J. Stachel in *PSA 1988*, vol. 2, edited by M. Forbes and A. Fine (East Lansing: Philosophy of Science Assoc., in press).

References

Adams, R. M. 1979. "Primitive Thisness and Primitive Identity." *Journal of Philosophy* 76:5–26.

Adler, R. J., Bazin, M. J., and Schiffer, M. 1975. *Introduction to General Relativity*. 2d. ed. New York: McGraw Hill.

Alexander, H. G., ed. 1984. *The Leibniz–Clarke Correspondence*. New York: Barnes and Noble.

Alexander, P. 1984/1985. "Incongruent Counterparts and Absolute Space." *Proceedings of the Aristotelian Society* 85:1–21.

Allison, H. 1983. *Kant's Transcendental Idealism*. New Haven: Yale University Press.

Anderson, J. L. 1967. *Principles of Relativity Physics*. New York: Academic Press.

Armstrong, D. 1983. *What Is a Law of Nature?* Cambridge: Cambridge University Press.

Barbour, J. B. 1974. "Relative-Distance Machian Theories." *Nature* 249:328–329. Erratum 250:606.

Barbour, J. B. 1975. "Forceless Machian Theories." *Nuovo Cimento* 26B:16–21.

Barbour, J. B. 1982. "Relational Concepts of Space and Time." *British Journal for the Philosophy of Science* 33:251–274.

Barbour, J. B. 1986. "Leibnizian Time, Machian Dynamics, and Quantum Gravity." In Penrose and Isham 1986.

Barbour, J. B., and Bertotti, B. 1977. "Gravity and Inertia in a Machian Framework." *Nuovo Cimento* 38B:1–27.

Barbour, J. B., and Bertotti, B. 1982. "Mach's Principle and the Structure of Dynamical Theories." *Proceedings of the Royal Society* (London) 382:295–306.

Beem, J. K. 1980. "Minkowki Space-Time Is Locally Extendible." *Communications in Mathematical Physics* 72:273–275.

Bennett, J. 1970. "The Difference between Right and Left." *American Philosophical Quarterly* 7:175–191.

Berkeley, G. 1710. *Principles of Human Knowledge*. In *Principles, Dialogues, and Philosophical Correspondence*, ed. and trans. by C. M. Turbayne. Indianapolis: Bobbs-Merrill, 1965.

Berkeley, G. 1712. *De motu*. In *Works of George Berkeley*, ed. and trans. by A. A. Luce and T. E. Jessop, vol. 4. London: T. Nelson, 1948–1957.

Bernstein, H. 1984. "Leibniz and Huygens on the 'Relativity' of Motion." *Studia Leibnitiana*, Sonderheft 13:85–101.

Bertotti, B., and Easthope, P. 1978. "The Equivalence Principle According to Mach." *International Journal of Theoretical Physics* 17:309–318.

Born, M. 1909. "Die Theorie des starren Elektrons in der Kinematik des Relativitätsprinzips." *Annalen der Physik* 30:1–56.

Born, M. 1910. "Zur kinematik des starren Körpers im System des Relativitätsprinzips." *Nachrichten von der Königlichen Gesellschaft der Wissenschaften zu Göttingen*, 161–179.

Broad, C. D. 1946. "Leibniz's Last Controversy with the Newtonians." *Theoria* 12:143–168.

Broad, C. D. 1978. *Kant: An Introduction*. Cambridge: Cambridge University Press.

Buchdahl, G. 1969. *Metaphysics and Philosophy of Science*. Cambridge: MIT Press.

Bunge, M., and Maynez, A. G. 1976. "A Relational Theory of Physical Space." *International Journal of Theoretical Physics* 15:961–972.

Buroker, J. V. 1981. *Space and Incongruence*. Dordrecht: D. Reidel.

Butterfield, J. 1984. "Relationism and Possible Worlds." *British Journal for the Philosophy of Science* 35:101–116.

Butterfield, J. 1989. "The Hole Story." *British Journal for the Philosophy of Science* 40:1–28.

Carter, B. 1973. "Black Hole Equilibrium States." In *Black Holes*, ed. by C. DeWitt and B. S. DeWitt. New York: Gordon and Breach.

Catton, P., and Solomon, G. 1988. "Uniqueness of Embedding and Space-Time Relationism." *Philosophy of Science* 55:280–291.

Clarke, C. J. S. 1976. "Space-Time Singularities." *Communications in Mathematical Physics* 49:17–23.

Clarke, S. 1738. *Works of Samuel Clarke*. 4 vols. London: J. and P. Knapton.

Couturat, L., ed. 1903. *Opuscules et fragments inédite de Leibniz*. Hildesheim: Georg Olms.

Cover, J. A., and Hartz, G. 1986. "Space and Time in the Leibnizian Metaphysic." *Noûs*, forthcoming.

Craig, W. 1956. "Replacement of Auxiliary Expressions." *Philosophical Review* 65:38–55.

Crawford, F. S., Cresti, M., Good, M. L., Gottstein, K., Lyman, E. M., Solmitz, F. T., Stevenson, M. L., and Ticko, H. K. 1957. "Detection of Parity Nonconservation in Λ Decay." *Physical Review* 108:1102–1103.

Degas, R. 1958. *Mechanics in the Seventeenth Century*. Neuchatel, Switzerland: Editions du Griffon.

Descartes, R. 1644. *Principles of Philosophy*. Trans. by V. R. Miller and R. P. Miller. Dordrecht: D. Reidel, 1984.

Des Coudres, T. 1889. "Über das Verhalten des lichtäthers Bewegungen der Erde." *Annalen der Physik und Chemie* 38:71–79.

Earman, J. 1970. "Who's Afraid of Absolute Space." *Australasian Journal of Philosophy* 48:287–317.

Earman, J. 1971. "Kant, Incongruous Counterparts, and the Nature of Space and Space-Time." *Ratio* 13:1–18.

Earman, J. 1978. "Perceptions and Relations in the Monadology." *Studia Leibnitiana* 9:212–230.

Earman, J. 1979. "Was Leibniz a Relationist?" In *Studies in Metaphysics*, ed. by P. Fench and H. Wettstein, Midwest Studies in Philosophy, vol. 4. Minneapolis: University of Minnesota Press.

Earman, J. 1986. *A Primer on Determinism*. Dordrecht: D. Reidel.

Earman, J. 1989. "Remarks on Relational Theories of Motion." *Canadian Journal of Philosophy* 19:83–87.

Earman, J., Glymour, C., and Rynasiewicz, R. 1983. "On Writing the History of Special Relativity." In *PSA 1982*, vol. 2, ed. by P. D. Asquith and T. Nicholes. East Lansing: Philosophy of Science Association.

Earman, J., and Norton, J. 1987. "What Price Space-Time Substantivalism? The Hole Story." *British Journal for the Philosophy of Science* 38:515–525.

Ehlers, J. 1973. "The Nature and Structure of Spacetime." In *The Physicist's Conception of Nature*, ed. by J. Mehra. Dordrecht: D. Reidel.

Einstein, A. 1905. "On the Electrodynamics of Moving Bodies." Reprinted in English translation in Perrett and Jeffrey 1952.

Einstein, A. 1914a. "Prinzipielles zur verallgemeinerten Relativitätstheorie." *Physikalische Zeitschrift* 15:176–180.

Einstein, A. 1914b. "Die formale Grundlage der allgemeinen Relativitätstheorie." *Preussische Akademie der Wissenschaften* (Berlin), Sitzungsberichte, pp. 831–839.

Einstein, A. 1916. "The Foundation of the General Theory of Relativity." Reprinted in English translation in Perrett and Jeffrey 1952.

Einstein, A. 1920. "Ether and the Theory of Relativity." In *Sidelights on Relativity*. New York: Dover, 1983.

Einstein, A. 1955. *The Meaning of Relativity*. 5th ed. Princeton: Princeton University Press.

Einstein, A. 1961. *Relativity: The Special and the General Theory*. New York: Bonanza Books.

Einstein, A., and Grossmann, M. 1913. "Entwurf einer verallgemeinerten Relativitätstheorie und einer Theorie der Gravitation." *Zeitschrift für Mathematik und Physik* 62:225–261.

Einstein, A., and Grossmann, M. 1914. "Kovarianzeigenschaften der Feldgleichungen der auf die verallgemeinerte Relativitätstheorie gegrundeten Gravitationstheorie." *Zeitschrift für Mathematik und Physik* 63:215–225.

Ellis, G. F. R., and Schmidt, B. G. 1977. "Singular Space-Times." *General Relativity and Gravitation* 8:915–953.

Erlichson, H. 1967. "The Leibniz–Clarke Controversy: Absolute versus Relative Space and Time." *American Journal of Physics* 35:89–98.

Field, H. 1980. *Science without Numbers*. Princeton: Princeton University Press.

Field, H. 1985. "Can We Dispense with Space-Time?" In *PSA 1984* vol. 2, ed. by P. D. Asquith and P. Kitcher. East Lansing: Philosophy of Science Association.

Feigl, H. 1953. "Notes on Causality." In *Readings in the Philosophy of Science*, ed. by H. Feigl and M. Brodbeck. New York: Appleton-Century-Crofts.

Föppl, A. 1897–1910. *Vorlesungen über technische Mechanik*. 6 vols. Leipzig: B. G. Teubner.

Friedman, M. 1973. "Relativity Principles, Absolute Objects, and Symmetry Groups." In *Space, Time, and Geometry*, ed. by P. Suppes. Dordrecht: D. Reidel.

Friedman, M. 1983. *Foundations of Space-Time Theories*. Princeton. Princeton University Press.

Gardner, G. H. F. 1952. "Right-Body Motions in Special Relativity." *Nature* 170:243.

Gardner, M. 1969. *The Ambidextrous Universe*. New York: Mentor Books.

Gerhardt, C. I., ed. 1849–1855. *G. W. Leibniz, Mathematische Schriften*. Vols. 1–2, Berlin: Asher. Vols. 3–7, Halle: H. W. Schmidt.

Gerhardt, C. I., ed. 1875–1890. *G. W. Leibniz, Philosophische Schriften*. 7 vols. Berlin: Weidmann.

Geroch, R. 1972. "Einstein Algebras." *Communications in Mathematical Physics* 26:271–275.

Geroch, R. 1977. "Prediction in General Relativity." In *Foundations of Space-Time Theories*, ed. by J. Earman, C. Glymour, and J. Stachel, Minnesota Studies in the Philosophy of Science, vol. 8. Minneapolis: University of Minnesota Press.

Geroch, R. 1978. *Relativity from A to B*. Chicago: University of Chicago Press.

Gillman, L., and Jerison, M. 1960. *Rings of Continuous Functions*. Princeton: Van Nostrand.

Glymour, C. 1980. *Theory and Evidence*. Princeton: Princeton University Press.

Griffiths, J. B. 1985. *The Theory of Classical Dynamics*. Cambridge: Cambridge University Press.

Grünbaum, A. 1973. *Philosophical Problems of Space and Time.* 2d ed. Dordrecht: D. Reidel.

Hacking, I. 1975. "The Identity of Indiscernibles." *Journal of Philosophy* 72:249–256.

Hall, A. R. 1980. *Philosophers at War.* Cambridge: Cambridge University Press.

Hall, A. R., and Hall, M. B., eds. 1962. *Unpublished Scientific Papers of Isaac Newton.* Cambridge: Cambridge University Press.

Hall, A. R., and Tilling, L., eds. 1975. *Correspondence of Isaac Newton.* Vol. 5. Cambridge: Cambridge University Press.

Harper, W. 1989. "Kant on Incongruent Counterparts." In *The Philosophy of Right and Left: Incongruent Counterparts and the Nature of Space*, ed. by J. Van Cleve. Dordrecht: Kluwer Academic, in press.

Harré, R. 1986. *Varieties of Realism.* Oxford: Basil Blackwell.

Havas, P. 1964. "Four-Dimensional Formulations of Newtonian Mechanics and Their Relation to the Special and the General Theory of Relativity." *Reviews of Modern Physics* 36:938–965.

Hawking, S. W., and Ellis, G. F. R. 1973. *The Large Scale Structure of Space-Time.* Cambridge: Cambridge University Press.

Hempel, C. G. 1965. *Aspects of Scientific Explanation.* New York: Free Press.

Heller, M., and Staruszkiewicz, A. 1975. "A Physicist's View on the Polemics between Leibniz and Clarke." *Organon* 11:205–213.

Herglotz, G. 1910. "Über den vom Standpunkt des Relativitätsprinzips aus als 'starr' zu bezeichnenden Körper." *Annalen der Physik* 31:393–415.

Hilbert, D. 1915. "Die Grundlagen der Physik (Erste Mitteilung)." *Nachrichten von der Königlichen Gesellschaft der Wissenschaften zu Göttingen*, pp. 395–407.

Hilbert, D. 1917. "Die Grundlagen der Physik (Zweite Mitteilung)." *Nachrichten von der Königlichen Gesellschaft der Wissenschaften zu Göttingen*, pp. 53–76.

Hill, R. N. 1967a. "Instantaneous Action-at-a-Distance in Classical Relativistic Mechanics." *Journal of Mathematical Physics* 8:201–220. Reprinted in Kerner 1972.

Hill, R. N. 1967b. "Canonical Formulation of Relativistic Mechanics." *Journal of Mathematical Physics* 8:1756–1773. Reprinted in Kerner 1972.

Holton, G. 1973. *Thematic Origins of Scientific Thought.* Cambridge: Harvard University Press.

Hood, C. G. 1970. "A Reformulation of Newtonian Dynamics." *American Journal of Physics* 38:438–442.

Hooker, C. A. 1971. "Relational Doctrines of Space and Time," *British Journal for the Philosophy of Science* 22:97–130.

Horwich, P. 1978. "On the Existence of Times, Space, and Space-Times." *Noûs* 12:396–419.

Horwich, P. 1987. *Asymmetries in Time.* Cambridge: MIT Press.

Huygens, C. 1888–1950. *Oeuvres complètes.* 22 vols. La Haye: Société Hollandaise des sciences.

Kant, I. 1768. "Concerning the Ultimate Foundation of the Differentiation of Regions in Space." In *Kant: Selected Pre-Critical Writings*, ed. and trans. by G. B. Kerferd and D. E. Walford. New York: Barnes and Noble, 1968.

Kant, I. 1770. "On the Form and Principles of the Sensible and Intelligible World." In *Kant's Inaugural Dissertation and Early Writings on Space*, trans. by J. Handyside. Westport, Conn.: Hyperion Press, 1979.

Kant, I. 1781. *Critique of Pure Reason.* Trans. by N. Kemp Smith. New York: St. Martin's Press, 1965.

Kant, I. 1783. *Prolegomena to Any Future Metaphysics.* Trans. J. W. Ellington. Indianapolis: Hackett, 1977.

Kant I. 1786. *Metaphysical Foundations of Natural Science.* Trans. by J. Ellington. Indianapolis: Bobbs Merrill, 1970.

Kant, I. 1790s. *What Real Progress Has Metaphysics Made in Germany since the Time of Leibniz and Wolff?* Trans. by T. Humphrey. New York: Abaris Books, 1983.

Kaplunovsky, V., and Weinstein, M. 1985. "Space-Time: Arena or Illusion?" *Physical Review* D31:1879–1898.

Kerner, E. H., ed. 1972. *Theory of Action-at-a-Distance in Relativistic Particle Dynamics.* New York: Gordon and Breach.

Koslow, A. 1967. *The Changeless Order.* New York: G. Braziller.

Koyré, A. 1965. *Newtonian Studies.* Chicago: University of Chicago Press.

Koyré, A., and Cohen, I. B. 1962. "Newton and the Leibniz–Clarke Correspondence." *Archives internationales d'histoire des sciences* 15:63–126.

Kuchar, K. 1981. "Gravitation, Geometry, and Nonrelativistic Quantum Theory." *Physical Review* D22:1285–1299.

Lacey, H. 1971. "The Philosophical Intelligibility of Absolute Space: A Study of Newtonian Argument." *British Journal for the Philosophy of Science* 21:317–342.

Lacey, H., and Anderson, E. 1980. "Spatial Ontology and Spatial Modalities." *Philosophical Studies* 38:261–285.

Lange, L. 1885. "Über die wissenschaftliche Fassung der Galilei'schen Beharrungsgesetzes." *Philosophische Studien* 2:266–297, 539–545.

Lange, L. 1886. *Die geschichtliche Entwicklung des Bewegungsbegriffes.* Leipzig: W. Englemann.

Lange, L. 1902. "Das Inertialsystem vor dem Forum der Naturforschung." *Philosophische Studien* 20:1–71.

Lariviere, B. 1987. "Leibnizian Relationism and the Problem of Inertia." *Canadian Journal of Philosophy* 17:437–448.

Laymon, R. 1978. "Newton's Bucket Experiment." *Journal of the History of Philosophy* 16:399–413.

Lee, T. D., and Yang, C. N. 1956. "Question of Parity Conservation in Weak Interactions." *Physical Review* 104:254–258.

Lewis, D. 1986. *On the Plurality of Worlds.* Oxford: B. Blackwell.

Loemker, L. E., ed. 1970. *Leibniz: Philosophical Papers and Letters.* Dordrecht: D. Reidel.

Lucas, J. R. 1984. *Space, Time, and Causality.* Oxford: Oxford University Press.

McGuire, J. E. 1976. "'Labyrinthus Continui': Leibniz on Substance, Activity, and Matter." In *Motion and Time, Space and Matter: Interrelations in the History and Philosophy of Science,* ed. by P. K. Machamer and R. G. Turnbull. Columbus: Ohio State University Press.

McGuire, J. E. 1978. "Existence, Actuality, and Necessity: Newton on Space and Time." *Annals of Science* 35:463–508.

Mach, E. 1883. *The Science of Mechanics.* 9th ed. London: Open Court, 1942.

McKinsey, J. C. C., and Suppes, P. 1955. "On the Notion of Invariance in Classical Mechanics." *British Journal for the Philosophy of Science* 5:290–302.

Malament, D. 1976. Review of Sklar 1976. *Journal of Philosophy* 73:306–323.

Malament, D. 1977. "Causal Theories of Time and the Conventionality of Simultaneity." *Noûs* 11:293–300.

Malament, D. 1978. Unpublished lecture notes.

Malament, D. 1985. "A Modest Remark about Reichenbach, Rotation, and General Relativity." *Philosophy of Science* 52:615–620.

Malament, D. 1986. "Why Space Must Be Euclidean." In *From Quarks to Quasars*, ed. by R. G. Colodny. Pittsburgh: University of Pittsburgh Press.

Manders, K. 1982. "On the Space-Time Ontology of Physical Theories." *Philosophy of Science* 49:575–590.

Maudlin, T. 1988. "Substances and Space-Time: What Aristotle Would Have Said to Einstein." Preprint.

Maxwell, J. C. 1877. *Matter and Motion*. New York: Dover, 1952.

Møller, C. 1952. *The Theory of Relativity*. Oxford: Oxford University Press.

Møller, C. 1972. *The Theory of Relativity*. 2d ed. Oxford: Oxford University Press.

Müller zum Hagen, H. 1972. "A New Physical Characterization of Stationary and Static Space-Times." *Proceedings of the Cambridge Philosophical Society* 71:381–389.

Mundy, B. 1983. "Relational Theories of Euclidean Space and Minkowski Space-Time." *Philosophy of Science* 50:205–226.

Mundy, B. 1986. "Embedding and Uniqueness in Relational Theories of Space." *Synthese* 67:383–390.

Mundy, B. 1987. "On Quantitative Relationist Theories." *Philosophy of Science*, forthcoming.

Nagata, J. 1974. *Modern General Topology*. Amsterdam: North-Holland.

Nagel, E. 1961. *The Structure of Science*. New York: Harcourt, Brace and World.

Nerlich, G. 1973. "Hands, Knees, and Absolute Space." *Journal of Philosophy* 70:337–351.

Nerlich, G. 1976. *The Shape of Space*. Cambridge: Cambridge University Press.

Newton, I. 1668? "De gravitatione." In *Unpublished Scientific Papers of Isaac Newton*, ed. A. R. Hall and M. B. Hall. Cambridge: Cambridge University Press, 1962.

Newton, I. 1684. "De motu corporum mediis regulariter cedentibus." In J. Herivel, *The Background to Newton's Principia*. Oxford: Clarendon Press, 1965.

Newton, I. 1729. *Mathematical Principles of Natural Philosophy*. Trans. by A. Motte and F. Cajori. Berkeley: University of California Press, 1962.

Noether, F. 1910. "Zur Kinematik des starren Körpers in Relativtheorie." *Annalen der Physik* 31:919–944.

Nomizu, K. 1956. *Lie Groups and Differential Geometry*. Tokyo: Herald Printing Co.

Norton, J. 1987. "Einstein, the Hole Argument, and the Objectivity of Space." In *Measurement, Realism, and Objectivity*, ed. by J. Forge. Dordrecht: D. Reidel.

Norton, J. 1988. "Coordinates and Covariance: Einstein's View of Space-Time and the Modern View." *Foundations of Physics*, forthcoming.

Ohanian, H. 1976. *Gravitation and Space-Time*. New York: W. W. Norton.

Ozsvath, I., and Schucking, E. 1962. "Finite Rotating Universe." *Nature* 193:1168–1169.

Penrose, R. 1968. "Structure of Space-Time." In *Battelle Rencontres*, ed. by C. M. De Witt and J. A. Wheeler. New York: W. A. Benjamin.

Penrose, R. 1971. "Angular Momentum: An Approach to Combinatorial Space-Time." In *Quantum Theory and Beyond*, ed. by T. Bastin. Cambridge: Cambridge University Press.

Penrose, R, 1987. "Newton, Quantum Theory, and Reality." In *Three Hundred Years of Gravitation*, ed. by S. W. Hawking and W. Israel. Cambridge: Cambridge University Press.

Penrose, R., and Isham, C. J., eds. 1986. *Quantum Concepts in Space and Time*. Oxford: Oxford University Press.

Perrett, W., and Jeffrey, G. B., eds. 1952. *The Principle of Relativity*. New York: Dover.

Pirani, F. A. E., and Williams, G. 1963. "Rigid Motions in a Gravitational Field." *Seminaire Janet* 5:8-01–8-16.

Poincaré, H. 1905. *Science and Hypothesis*. New York: Dover, 1952.

Popper, K. 1953. "A Note on Berkely as a Precursor of Mach." *British Journal for the Philosophy of Science* 4:26–36. Reprinted as "A Note on Berkeley as a Precursor of Mach and Einstein" in *Conjectures and Refutations*, 3d ed. London: Routledge and Kegan Paul.

Raine, D. J. 1981. "Mach's Principle and Space-Time Structure." *Reports on Progress in Physics* 44:1151–1195.

Reichenbach, H. 1924. "The Theory of Motion according to Newton, Leibniz, and Huyghens." Reprinted in *Modern Philosophy of Science*, ed. and trans. by M. Reichenbach. London: Routledge and Kegan Paul, 1959.

Reichenbach, H. 1957. *Space and Time*. New York: Dover.

Remnant, P. 1963. "Incongruent Counterparts and Absolute Space." *Mind* 72:393–399.

Rosen, G. 1972. "Galilean Invariance and the General Covariance of Nonrelativistic Laws." *American Journal of Physics* 40:683–687.

Rynasiewicz, R. A. 1986. "The Universality of Laws in Space and Time." In *PSA 1986*, vol. 1. East Lansing: Philosophy of Science Association.

Sakurai, J. J. 1964. *Invariance Principles and Elementary Particles*. Princeton: Princeton University Press.

Salmon, N. 1981. *Reference and Essence*. Princeton: Princeton University Press.

Salmon, W. 1984. *Scientific Explanation and the Causal Structure of the World*. Princeton: Princeton University Press.

Salvioli, S. E. 1972. "On the Theory of Geometric Objects." *Journal of Differential Geometry* 7:257–278.

Schouten, J. A. 1920. "Die relative und absolute Bewegung bei Huygens." *Jahresbericht der deutschen Mathematiker-Vereinigung* 29:136–144.

Schouten, J. A., and Haantjes, J. 1936. "On the Theory of the Geometric Object." *Proceedings of the London Mathematical Society* 42:356–376.

Shankland, R. S. 1964. "The Michelson–Morley Experiment." *American Journal of Physics* 32:16–35.

Shankland, R. S., McCuskey, S. W., Leone, F. C., and Kuerti, G. 1952. "New Analysis of the Interferometer Observations of Dayton C. Miller." *Reviews of Modern Physics* 27:167–178.

Shapin, S. 1981. "Of Gods and Kings: Natural Philosophy and Politics in the Leibniz–Clarke Disputes." *Isis* 72:187–215.

Sklar, L. 1974. "Incongruous Counterparts, Intrinsic Features, and the Substantivality of Space." *Journal of Philosophy* 71:227–290.

Sklar, L. 1976. *Space, Time, and Space-Time*. Berkeley: University of California Press.

Stachel, J. 1986. "What a Physicist Can Learn from the Discovery of General Relativity." In *Proceedings of the Fourth Marcel Grossmann Meeting on Recent Developments in General Relativity*, ed. by R. Ruffini. Amsterdam: North-Holland.

Stein H. 1967. "Newtonian Space-Time." *Texas Quarterly* 10:174–200.

Stein, H. 1977. "Some Pre-History of General Relativity." In *Foundations of Space-Time Theories*, ed. by J. Earman, C. Glymour, and J. Stachel, Minnesota Studies in the Philosophy of Science, vol. 8. Minneapolis: University of Minnesota Press.

Stewart, L. 1981. "Samuel Clarke, Newtonianism, and the Factions of Post-Revolutionary England." *Journal of the History of Ideas* 42:53–72.

Suppes, P. 1972. "Some Open Problems in the Philosophy of Space and Time." *Synthese* 24:298–316.

Synge, J. L. 1952a. "Gardner's Hypothesis and the Michelson–Morley Experiment." *Nature* 170:243–244.

Synge, J. L. 1952b. "Effects of Acceleration in the Michelson–Morley Experiment." *Scientific Proceedings of the Royal Dublin Society* 26:45–54.

Teller, P. 1987. "Space-Time as a Physical Quantity." In *Kelvin's Baltimore Lectures and Modern Theoretical Physics*, ed. by P. Achinstein and R. Kagon. Cambridge: MIT Press.

Torretti, R. 1983. *Relativity and Geometry*. Oxford: Pergamon Press.

Trautman, A. 1965. "Foundations and Current Problems of General Relativity." In *Lectures on General Relativity*, ed. by S. Deser and K. W. Ford. Englewood Cliffs, N.J.: Prentice-Hall.

Trautman, A. 1966. "Comparison of Newtonian and Relativistic Theories of Space-Time." In *Perspectives in Geometry and Relativity*, ed. by B. Hoffmann. Bloomington: Indiana University Press.

Trouton, F. T. 1902. "The Results of an Electrical Experiment, Involving the Relative Motion of the Earth and Ether." *Transactions of the Royal Dublin Society* 7:379–384.

Trouton, F. T., and Noble, H. R. 1904. "The Mechanical Forces Action on a Charged Electric Condensor Moving through Space." *Philosophical Transactions of the Royal Society* (London) 202:165–181.

Truesdell, C. 1977. *A First Course in Rational Continuum Mechanics*. New York: Academic Press.

Van Cleve, J. 1987. "Right, Left, and the Fourth Dimension." *Philosophical Review* 96:33–68.

Van Dam, H., and Wigner, E. P. 1966. "Instantaneous and Asymptotic Conservation Laws for Classical Relativistic Mechanics of Interacting Point Particles." *Physical Review* 142:883–843. Reprinted in Kerner 1972.

Van Fraassen, B. C. 1970. *An Introduction to the Philosophy of Time and Space*. New York: Random House.

Van Fraassen, B. C. 1980. *The Scientific Image*. Oxford: Oxford University Press.

Wald, R. M. 1984. *General Relativity*. Chicago: University of Chicago Press.

Wald, R. M. 1986. "Correlations and Causality in Quantum Field Theory." In Penrose and Isham 1986.

Walker, R. C. S. 1978. *Kant*. London: Routledge and Kegan Paul.

Westfall, R. S. 1971. *Force in Newton's Physics*, New York: American Elsevier.

Weyl, H. 1966. *The Philosophy of Mathematics and Natural Science*. Princeton: Princeton University Press.

Whitrow, G. J., and Morduch, G. E. 1965. "Relativistic Theories of Gravitation." In *Vistas in Astronomy*, vol. 6, ed. by A. Beer. Elmsford, N.Y.: Pergamon Press.

Winterbourne, A. T. 1981. "On the Metaphysics of Leibnizian Space and Time." *Studies in the History and Philosophy of Science* 13:201–214.

Winterbourne, A. T. 1982. "Incongruent Counterparts and the Intuitive Nature of Space." *Auslegung* 9:85–98.

Witten, L. 1988a. "Topological Gravity." *Physics Letters* B206:601–606.

Witten, L. 1988b. "Topological Quantum Field Theory." *Communications in Mathematical Physics* 117:353–386.

Wolff, R. P. 1969. *Kant's Theory of Mental Activity.* Cambridge: Harvard University Press.

Zanstra, H. 1924. "A Study of Relative Motion in Connection with Classical Mechanics." *Physical Review* 23:528–545.

Index

LIBRARY
OF
MOUNT ST. MARY'S
COLLEGE